Crystal Research and Technology

7

Journal of Experimental and Industrial Crystallography

Zeitschrift für experimentelle und technische Kristallographie

Volume 22
1987
Number 7 July

THEORIA CUM PRAXI

Akademie-Verlag · Berlin

ISSN 0232-1300 Cryst. Res. Technol., Berlin 22 (1987) 7, 867–986, K113–K123

Directions to the author

In the journal "Crystal Research and Technology – Kristall und Technik" as a rule the following categories of papers will be printed: original papers, review articles, short notes, and book reviews, **preferentially in English**. Complete information necessary for type-setting and a "Short Title" should be added.

The manuscript should not exceed 10 pages and be type-written on one side only in the format A4 (21 × 29.7 cm) on white ink-proof paper, with a two-lined distance and a left-hand free edge of 3 cm, and submitted in 2 copies. All sheets of paper belonging to the contribution are continuously to be numbered.

The head of the manuscript should be arranged in this manner: initials and name of the author; institution, and title of the paper. Then two abstracts should follow, one in the original language and the second in German, English, French, or Russian, respectively.

Formulae, signs and symbols, letters etc. being wanting on the type-writer, should be distinctly written with penholder. 0 as zero is to be underlined, the capital O is not. Greek letters should be underlined with red ink.

Paragraphs should be marked by indenting of at least 3 touches.

Concerning the marking of the manuscript should be noted:
headings should be underlined, sub-headings not, and indented by 9 touches. The titles of tables are not to be underlined. Other markings in the text will be done by the editorial staff. They mean:

light-red ∿∿∿∿∿∿∿ = small capitals	light-red	perpendicular straight red line on the left edge = small printing
blue ∿∿∿∿∿∿∿ = in italics		

Tables are to be added on separate sheets of paper at the end of the contribution.

Foot notes should be continuously written on a separate sheet of paper.

Figures (Fig.) – drawings and photographs – should be not more than 8 and submitted (2 copies) in a special envelope: photographs as white high-gloss prints (inscription on a transparent covering sheet) and drawings as distinct sketches. On the back of the submitted figures the reduction scale in tenth (9/10, 8/10, 5/10 e.g.), author's name, short title of the paper, and the number of the figure should be given.

Captions for the figures should be submitted on a separate sheet of paper.

References should be written without numbers in alphabetical order of the names at the end of the contribution. Cited names in the text stand for the references. The references should contain the following items, for example:

for books: RAST, K.: Physikalisch-chemische Rechnungen, Berlin 1958
for journals: KRAUSE, H. N., MÜLLER, K.: Cryst. Res. Technol. **16**, 70 (1981)

Add, please, the short title, about 6 keywords, and the academic grades of the authors.

The authors will be sent two galley proofs of review articles and original papers. Proof reading of short notes and book reviews will be done by the editorial staff.

For each of the papers a total of 50 reprints is available free of charge. Requests for additional maximum 200 reprints (against payment) should be given together with the submitted manuscript.

7

Crystal Research and Technology

Founded 1966 as
Kristall und Technik

Journal of Experimental and Industrial Crystallography

VOLUME 22 · 1987 · NUMBER 7

Akademie-Verlag · Berlin

Papers should be directed to the editor's office
Scharnhorststraße 20, DDR-7030 Leipzig,
or to the editor Prof. Dr. Hermann Neels,
Zeititzer Weg 25, DDR-7254 Machern,
or to one of the members of the Advisory Board.

Orders can be sent

- in the GDR:
  to the Postzeitungsvertrieb or to the Akademie-Verlag Berlin, Leipziger Straße 3–4, DDR-1086 Berlin;
- in the other socialist countries: to a book-shop for foreign language literature or to the competent news-distributing agency;
- in the FRG and Berlin (West): to a book-shop or to the wholesale distributing agency Kunst und Wissen, Erich Bieber OHG, Wilhelmstr. 4–6, D-7000 Stuttgart 1;
- in the other Western European countries: to Kunst und Wissen, Erich Bieber GmbH, General Wille-Str. 4, CH-8002 Zürich;
- in USA and Canada to the VCH Publishers, Inc., 303 NW 12th Avenue, Deerfield Beach, FL 33442-1788, USA;
- in other countries: to the international book- and journal-selling trade, to Buchexport, Volkseigener Außenhandelsbetrieb der DDR, Postfach 160, DDR-7010 Leipzig, or to the Akademie-Verlag Berlin, Leipziger Straße 3–4, DDR-1086 Berlin.

Journal „Crystal Research and Technology"

Herausgeber: Im Auftrage des Akademie-Verlages Berlin herausgegeben von Prof. Dr. Hermann Neels.

Verlag: Akademie-Verlag Berlin, Leipziger Str. 3–4, DDR-1086 Berlin; Fernruf: 2236221 oder 2236229; Telex-Nr.: 114420; Bank: Staatsbank der DDR, Berlin, Kto.-Nr. 6836-26-20712.

Chefredakteur: Prof. Dr. Hermann Neels, Zeititzer Weg 25, DDR-7254 Machern; Fernruf: Brandis 3442.
Anschrift der Redaktion: Scharnhorststr. 20, DDR-7030 Leipzig; Telefon: 310502.

Veröffentlicht unter der Lizenznummer 1526 des Presseamtes beim Vorsitzenden des Ministerrates der Deutschen Demokratischen Republik.

Gesamtherstellung: VEB Druckerei „Thomas Müntzer", DDR-5820 Bad Langensalza.

Erscheinungsweise: Das Journal „Crystal Research and Technology" erscheint monatlich. Die 12 Hefte eines Jahrganges bilden einen Band. Bezugspreis je Band 480,– DM zuzüglich Versandspesen; Preis je Heft 40,– DM. Der gültige Jahresbezugspreis für die DDR ist der Postzeitungsliste zu entnehmen.

Bestellnummer dieses Heftes: 1082/22/7.

AN (EDV) 35830
02800

Cryst. Res. Technol.	22	1987	7	869–875

B. LORENZ

Original Papers

Akademie der Wissenschaften der DDR, Zentralinstitut für Physik der Erde, Potsdam

Grain Size Distribution in the Kolmogorov Model for Nucleation and Growth: Heterogeneous versus Homogeneous Nucleation

A theory of grain size distribution in nucleation and growth reactions described by the KOLMOGOROV model is developed. The distribution of agglomerates of growing and impinging nuclei is explicitely calculated for the two cases of heterogeneous and homogeneous nucleation. The results are compared and discussed with respect to the investigation of the nucleation mechanism in crystallization processes.

Es wird eine Theorie der Korngrößenverteilung entwickelt für Keimbildungs- und Wachstumsprozesse, die durch das KOLMOGOROV-Modell beschrieben werden. Für die zwei Fälle der heterogenen und der homogenen Keimbildung werden die Verteilungen der Agglomerate ineinanderwachsender Keime explizit berechnet. Die Ergebnisse werden verglichen und bezüglich ihrer Anwendung auf die Untersuchung des Keimbildungsmechanismus in Kristallisationsprozessen diskutiert.

1. Introduction

The kinetics of first order phase transitions has been a subject of increasing attention (see e.g. BINDER, STAUFFER). Thereby, the time development of the stable phase within the metastable initial phase of a thermodynamic system proceeds in two steps. At the very beginning of the reaction nuclei of the stable phase are formed due to statistical fluctuations. These nuclei may grow or decompose again depending on their size compared with the critical size. The kinetics of this transient phase is mainly influenced by thermodynamic fluctuations and is best described by a stochastic master equation until the steady state nucleation rate for critical nuclei is attained. In the steady state critical nuclei are created with mean nucleation rate $I(t)$ and grow with linear velocity $v = dR/dt$ (for simplicity we consider isotropic conditions where the nuclei are spherically shaped with radius R). The stage of steady state nucleation and growth is characterized by a deterministic development of the system and is frequently described by the Kolmogorov model (KOLMOROGOV; AVRAMI; JOHNSON, MEHL). According to the physical mechanisms of nucleation and growth the kinetic process is determined by the two quantities $I(t)$ and v. For interface controlled phase transformations the growth rate v is constant (see e.g. CHRISTIAN). If two growing particles impinge they continue to grow at the same rate, except at their common interface.

The nucleation rate $I(t)$ can be an arbitrary function of time t. We will consider the two limit cases,

$$I(t) = n_0\delta(t)\,, \qquad \text{heterogeneous nucleation,}$$

and

$$I(t) = I_0 = \text{constant}\,, \qquad \text{homogeneous nucleation.}$$

n_0 is the density of nuclei per unit volume and $\delta(t)$ is the Dirac δ-function.

In many cases the transient stage of the transformation is very short and can be neglected. Furthermore, the size of the critical nuclei is small and the growth process may be considered as starting with zero radius. With these assumptions the Kolmogorov model yields an adequate description of the transformation kinetics.

Only few exact results of the KOLMOGOROV model are known. The volume fraction of the transformed phase is given by the JOHNSON-MEHL-AVRAMI equation

$$x(t) = 1 - \exp\{-\tfrac{4}{3}\pi v^3 \int_0^t I(t')\,(t-t')^3\,dt'\} \tag{1}$$

and the total number of nuclei at time t follows from (1),

$$z(t) = \int_0^t I\,(t')(1 - x(t'))\,dt'\,. \tag{2}$$

The kinetics of the one-dimensional model has been treated rigorously by SEKIMOTO and by AXE and YAMADA.

During their growth neighbouring particles impinge and form larger aggregates which we denote as grains. Grains consisting of a given number of nuclei differ in their shape and volume. The time dependent grain size distribution is a characteristic function which strongly depends on the nucleation process. In this paper a general formalism for the calculation of the grain size distribution is given (Sect. 2) and applied to the cases of heterogeneous and homogeneous nucleation (Sect. 3).

2. Grain size distribution

We calculate the grain size distribution function $W(R) = \sum_j W_i(R)$ by the successive calculation of the partial distribution functions $W_j(R)$ ($j = 1, 2, 3 \ldots$) for grains consisting of j nuclei. To characterize the size of a grain we introduce an effective radius R by $\frac{4}{3}\pi R^3 = V_g$, where V_g denotes the volume of the grain. By this definition $W(R)\,dR$ is the number of grains per unit volume with effective radii between R and $R + dR$.

The problem results in the calculation of the volume distribution at time t in an ensemble of grains built up by nuclei of radius $r_i = v(t - t_i)$. t_i is the creation time of the i-th nucleus. We assume that the nuclei grow spherically with constant velocity. Depending on the nucleation process there results a distribution of radii of the growing nuclei $n(r, t)$ which is the basic function for the calculation of $W(R)$. In order to treat nuclei of different radii as statistically independent we allow for the creation and growth of phantom nuclei in the already transformed phase. This procedure corresponds to the extended volume concept applied to the derivation of (1) (CHRISTIAN). In the calculation of the grain size distribution the phantom nuclei have to be excluded.

The distribution $n(r, t)$ is then given by

$$n(r, t) = \frac{1}{v}\, I\left(t - \frac{r}{v}\right) \theta\left(t - \frac{r}{v}\right). \tag{3}$$

The step function $\theta(x)$ limits the radii of the nuclei to be less than the largest possible radius vt. In the following we consider a fixed time t and suppress the time argument in $n(r, t)$.

2.1. Isolated nuclei, $j = 1$

The distribution of nuclei which show no impingement with other ones is written as

$$W_1(R)\,dR = n(R)\,W^0(R)\,dR\,. \tag{4}$$

$W^0(R)$ is the probability that a growing nucleus with radius R does not impinge with any other nucleus and is calculated by simple statistical considerations for an infinitely large system,

$$W^0(R) = \exp\left\{-\tfrac{4}{3}\pi\int_0^R[(R+r)^3-(R-r)^3]\,n(r)\,dr - \right.$$
$$\left. -\tfrac{4}{3}\pi\int_R^\infty(R+r)^3\,n(r)\,dr\right\}. \tag{5}$$

The exponentials of the first and second integral in (5) represent the probabilities that no further nuclei with radii $r < R$ and $r > R$, respectively, impinge with the nucleus of radius R. Thereby, phantom nuclei are excluded.

Let us note that the volume fraction of the untransformed phase is given by $1 - x = W^0(R = 0) = \exp\{-\tfrac{4}{3}\pi\int_0^\infty r^3n(r)\,dr\}$. Substituting $r = v(t - t')$ equation (1) is derived.

2.2. Two connected nuclei, $j = 2$

The probability distribution of $j = 2$ cluster is given by

$$W_2(R_1, R_2, y)\,dR_1\,dR_2\,dy = \tilde{W}_2(R_1, R_2, y)\,W_2^0(R_1, R_2, y)\,dR_1\,dR_2\,dy\,. \tag{6}$$

R_1, R_2, and y are the radii of the two nuclei and their overlap distance, respectively (Fig. 1). The probability W_2^0 that the cluster has no contact to other nuclei is calculated analogous to the former case $j = 1$. $\tilde{W}_2(R_1, R_2, y)$ denotes the density of pairs of nuclei with radii R_1, R_2 and overlap y and is given by

$$\tilde{W}_2(R_1, R_2, y) = n(R_1)\,n(R_2)\,2\pi(R_1 + R_2 - y)^2\,. \tag{7}$$

The volume of that cluster is calculated as

$$V_2(R_1, R_2, y) = \frac{4}{3}\pi\left\{R_1^3 + R_2^3 - \left(\frac{y}{4}\right)^2[a_2^2(3R_1 - a_2y/2) + a_1^2(3R_2 - a_1y/2)]\right\}, \tag{8}$$

$$a_i = (2R_i - y)/(R_1 + R_2 - y)\,.$$

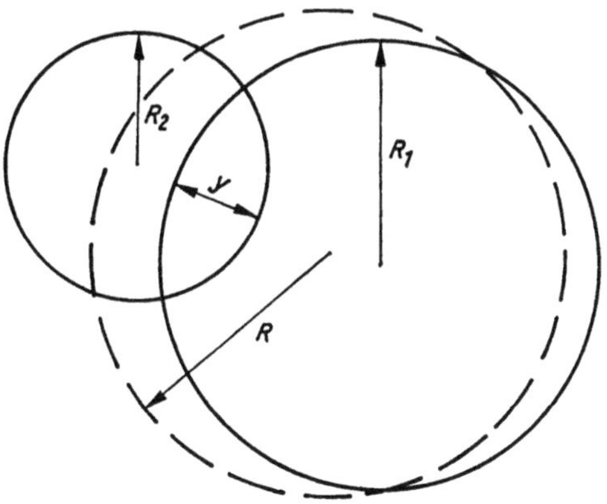

Fig. 1. Cluster of two nuclei with radii R_1, R_2 and overlap y. The effective grain with radius R is shown by the dashed line

The three-parameter distribution (6) is reduced to a one-parameter probability distribution by the summation over all cluster with equal volume. Introducing the effective radius R defined above we get

$$W_2(R)\,dR = \int W_2(R_1, R_2, y)\,\delta\left(R - \left[\frac{3}{4\pi} V_2(R_1, R_2, y)\right]^{1/3}\right) dR_1\,dR_2\,dy\,dR\,. \tag{9}$$

2.3. Larger aggregates, $j > 2$

The partial distribution functions W_j for $j = 1, 2$ are given rigorously by (4), (9). However, for $j > 2$ approximations have to be introduced. For the calculation of W_3 we substitute the $j = 2$ cluster by spherically shaped nuclei with effective radius R and density $n_2(R)$ which is defined by

$$W_2(R)\,dR = n_2(R)\,W^0(R)\,dR\,. \tag{10}$$

$W_2(R)$ and $W^0(R)$ are given by (9) and (5), respectively. The distribution of the $j = 2$ cluster is thus approximated by that of spherical nuclei with the renormalized density $n_2(R)$ and the algorithm described above can now be applied to the calculation of W_3. Repeating this procedure we get a recursion relation for W_j

$$W_j(R)\,dR = \int n_{j-1}(R_1)\,n(R_2)\,2\pi(R_1 + R_2 - y)^2\,W_2^0(R_1, R_2, y)\,\times$$

$$\times\,\delta\left(R - \left[\frac{3}{4\pi} V_2(R_1, R_2, y)\right]^{1/3}\right) dR_1\,dR_2\,dy\,dR\,, \tag{11}$$

$$W_{j-1}(R)\,dR = n_{j-1}(R)\,W^0(R)\,dR\,.$$

3. Results and discussion

With the partial grain size distribution functions (11) some characteristic quantities can be calculated. The total number of grains per unit volume consisting of j nuclei and the net volume fraction of these grains are given by

$$z_j = \int W_j(R)\,dR\,, \qquad x_j = \int \tfrac{4}{3}\pi R^3 W_j(R)\,dR\,. \tag{12}$$

In the following we consider the two cases of heterogeneous and homogeneous nucleation as defined in Section 1 and calculate the quantities (12) and the grain size distribution function $W(R)$.

3.1. Heterogeneous nucleation

Heterogeneous nucleation takes place if there exist preferred nucleation sites due to impurities or grain boundaries. In this case all nuclei are created at the beginning of the kinetic process and have the same radius $R_0 = vt$ during their growth. The density distribution is given by

$$n(R) = n_0\delta(R - R_0)\,. \tag{13}$$

It is convenient to express the relevant quantities in terms of the dimensionless length and time variables $\nu = (4\pi n_0/3)^{1/3} R$, $\nu_0 = (4\pi n_0/3)^{1/3} vt$. With (13) we get

$$1 - x = \exp(-\nu_0^3)\,, \qquad z_1 = n_0 \exp(-8\nu_0^3)\,, \qquad x_1 = \nu_0^3 \exp(-8\nu_0^3)\,,$$

$$W_1(R)\,dR = W_1(\nu)\,d\nu = n_0\delta(\nu - \nu_0)\exp(-8\nu_0^3)\,d\nu\,. \tag{14}$$

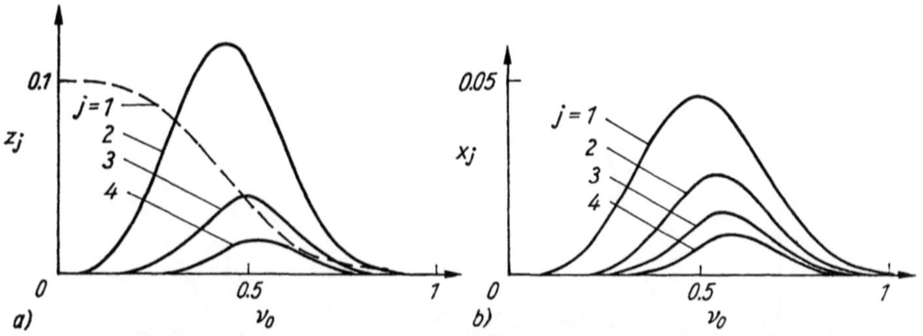

Fig. 2. Densities z_j (a) and volume fractions x_j (b) for grains containing 1 to 4 nuclei: heterogeneous nucleation .Note that z_1 (dashed line) is reduced by a factor of ten in the figure

The result (14) has also been obtained by MARKWORTH. z_j, x_j for $j > 1$ are calculated from (11), (12) with (13) and are shown in Figure 2 ($j = 1, \ldots, 4$). The density of single nuclei decreases from its initial value n_0 at $\nu_0 = 0$ due to the impingement of nuclei for $\nu_0 > 0$. The density of grains consisting of two or more nuclei increases with increasing ν_0 until it reaches a maximum and decreases again due to the aggregation of smaller cluster to larger ones (Fig. 2a). The characteristic maximum is shifted to larger times with increasing number of nuclei per grain. A similar behaviour is found in the time dependence of x_j (Fig. 1b).

3.2. Homogeneous nucleation

In the absence of preferred nucleation sites nuclei are created by thermal fluctuations in the metastable initial phase at constant nucleation rate I_0. The density distribution is obtained as

$$n(R) = \frac{1}{v} I_0 \theta(R_0 - R) . \tag{15}$$

$R_0 = vt$ represents the largest possible radius of the nuclei at time t. Introducing the reduced quantities $\mu = ((\pi/3) I_0/v)^{1/4} R$ and $\mu_0 = ((\pi/3) I_0/v)^{1/4} vt$ we get

$$1 - x = \exp(-\mu_0^4) , \qquad z_1 = \sqrt[4]{3/\pi}\, (I_0/v)^{3/4} \int_0^{\mu_0} W^0(\mu)\, d\mu ,$$

$$x_1 = 4 \int_0^{\mu_0} \mu^3 W^0(\mu)\, d\mu ,$$

$$W_1(R)\, dR = W_1(\mu)\, d\mu = \sqrt[4]{3/\pi}\, (I_0/v)^{3/4}\, W^0(\mu)\, \theta(\mu_0 - \mu)\, d\mu , \tag{16}$$

with

$$W^0(\mu) = \exp(-(\mu + \mu_0)^4 + 2\mu^4) .$$

Inserting (15) into (11) the corresponding quantities for $j > 1$ are calculated and shown in Figure 3. In contrast to the heterogeneous nucleation the density z_1 first increases due to permanent creation of nuclei, passes a maximum and decreases at larger μ_0 because of impingement of the growing nuclei. The densities of larger aggregates z_j ($j > 1$) (Fig. 3a) and the volume fractions x_j (Fig. 3b) look like the corresponding quantities in the case of homogeneous nucleation.

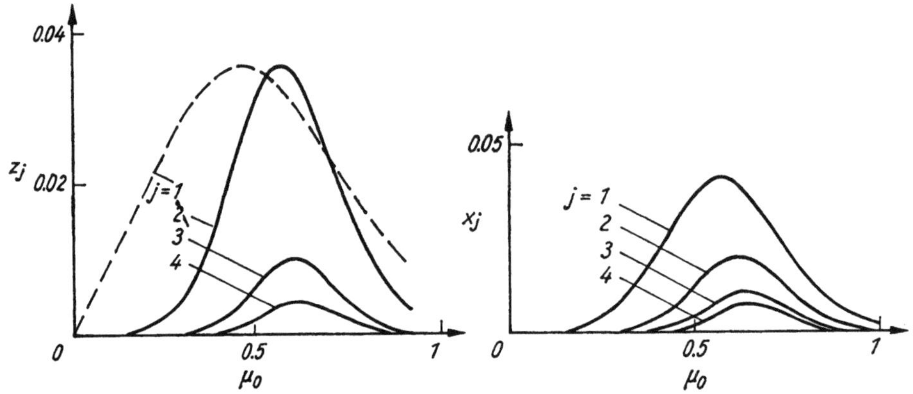

Fig. 3. Same as Figure 2: homogeneous nucleation

3.3. Grain size distributions

In Figure 4 the grain size distributions $W(R) = \sum_j W_j(R)$ for heterogeneous and homogeneous nucleation are shown at a reaction time corresponding to a transformed volume fraction of $x = 0.1$. Obviously, the distribution functions show the most striking differences. In the case of heterogeneous nucleation (Fig. 4a) at least three sharp peaks are resolved. The first one resulting from $W_1(R)$ is a δ-function singularity at $R = R_0$. The second peak is mainly due to $W_2(R)$ and shows a divergence at $R = R_0 \sqrt[3]{2}$. The partial distributions for larger cluster ($j < 2$) give rise to some further peaks with decreasing height.

The grain size distribution for homogeneous nucleation is a continuous function except of a remarkable step at $R = R_0$, the largest radius for the nuclei (Fig. 4b). For $R > R_0$ no single nuclei contribute to $W(R)$ and it decreases discontinuously. Furthermore, grains with arbitrary small size are present in a considerable number.

It is worth noting that the grain size distribution defined in this paper loses its sence if the largest grain reaches the size of the system (or infinity for an infinit system). Therefore, we restrict our considerations to transformed volume fractions $x < x_c$

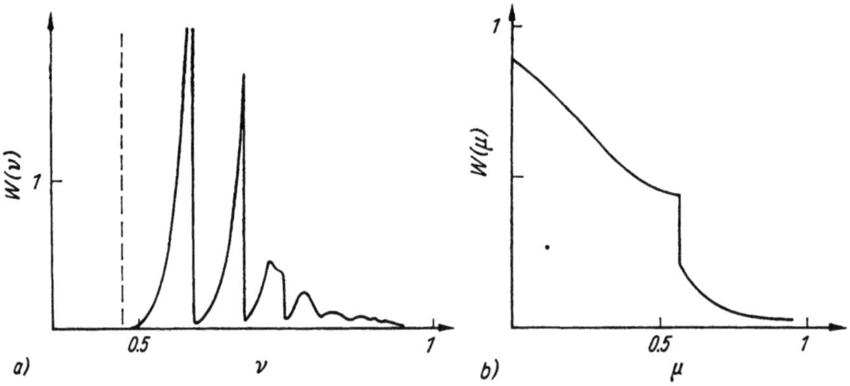

Fig. 4. Grain size distribution functions for: a) heterogeneous nucleation ($W(\nu)$ in units of n_0). The δ-function singularity is marked by the dashed line; b) homogeneous nucleation ($W(\mu)$ in units of $[3/\pi(I_0/v)^3]^{1/4}$)

where x_c determines the percolation threshold above which an infinitely large cluster is formed (for heterogeneous nucleation e.g. x_c is about 0.3). Above x_c the transformed phase is characterized by a polycrystalline aggregate and its description by the distribution (11) is inadequate.

The experimental determination of the dominant nucleation mechanism in a nucleation and growth reaction is a complicated task. Usually, the time exponent in the exponential of the JOHNSON-MEHL-AVRAMI equation (1) is determined by careful measurements of the transformed volume fraction $x(t)$ in order to select the physical nucleation mechanism (CHRISTIAN). As demonstrated above the grain size distribution function is very sensitive to the underlying nucleation scheme and yields important information on the physics of nucleation and growth processes.

The author wishes to thank M. Riedel and I. Orgzall for valuable discussions.

References

AVRAMI, M.: J. Chem. Phys. 7, 1103 (1939); 8, 212 (1940); 9, 177 (1941)
AXE, J. D., YAMADA, Y.: Phys. Rev. B 34, 1593 (1986)
BINDER, K., STAUFFER, D.: Adv. Phys. 25, 343 (1976)
CHRISTIAN, J. W.: The Theory of Transformations in Metals and Alloys, Oxford 1965
JOHNSON, W. A., MEHL, R. F.: Trans. Am. Inst. Mining Metall. Engrs. 135, 416 (1939)
KOLMOGOROV, A. N.: Bull. Acad. Sci. USSR, Phys. Ser. 3, 355 (1938)
MARKWORTH, A. J.: Scripta Metall. 18, 1309 (1984)
SEKIMOTO, K.: Physica A 125, 261 (1984)

(Received December 9, 1986)

Author's adress:

Dr. sc. B. LORENZ

Zentralinstitut für Physik der Erde, Bereich V
Akademie der Wissenschaften der DDR
Telegrafenberg
Potsdam
DDR-1500

Cryst. Res. Technol.	22	1987	7	876

Book Review

B. Feuerbacher, H. Hamacher, R. J. Naumann (eds)

Materials Sciences in Space

Springer Verlag Berlin, Heidelberg, New York, Tokyo 1986, 496 Seiten mit 199 Abbildungen und 21 Tabellen, Sachwörterverzeichnis, Preis 138,– DM, ISBN 3-540-16358-1

Die Spezialliteratur zu materialwissenschaftlichen Experimenten unter Schwerelosigkeit, insbesondere in Weltraumlaboratorien, aber auch in ballistischen Forschungsraketen und Falltürmen ist bereits beträchtlich angewachsen. Deshalb ist eine zusammenfassende Darstellung sehr zu begrüßen. Dadurch wird ein Überblick über Besonderheiten, Sinn und Entwicklungslinien materialwissenschaftlicher Forschungen unter Mikrogravitation vermittelt.

Für dieses Anliegen wurden 19 namhafte internationale Autoren verschiedener Forschungsprogramme und -laboratorien der USA und Westeuropas (NASA, ESA, DFVLR, u. a.) gewonnen, die in 19 in sich abgeschlossenen Abschnitten mit Literaturübersichten spezielle Fragestellungen darstellen.

Im Mittelpunkt steht der Phasenübergang flüssig-fest, explizite die gerichtete Erstarrung mit den technischen Hintergründen der Einkristallzüchtung und Metallurgie von Gefügestrukturen. Behandelt wird aber auch die Glasbildung und Separation biologischer Zellen und Proteine. Der erste Teil ("Microgravity Materials Sciences") führt in die Problematik der Werkstofforschung unter reduzierter Gravitation, in die Motivation der Forschungsprogramme ein, zeigt die historische Entwicklung der Weltraumlabortechnik und von Parabelflügen und vermittelt theoretische Grundkenntnisse zur Berechnung der Absolut- und Relativgrößen der Mikrogravitation als Funktion der Ballistikparameter und bordabhängigen Beschleunigungsschwankungen. Der zweite Teil („Physical Phenomena") behandelt die Theorie der physikalischen Erscheinungen unter Mikrogravitation, wie fehlende Volumenkonvektion, Diffusion und Atomtransport, das spezielle Verhalten der Kapillarität, Benetzung, Erstarrung, Unterkühlung und Keimbildung sowie die Vorteile der Mikrogravitation für die Untersuchung kritischer Punkte. Besonders günstige Bedingungen existieren zur Untersuchung von Segregationserscheinungen, der morphologischen Stabilität und Instabilität kristallisierender Phasengrenzen sowie des homogenen Keimbildungsvorganges und amorpher Erstarrungen. Im dritten Teil („Experimental Hardware") werden apparative Lösungen zur Realisierung der Experimente unter den Bedingungen der Mikrogravitation vorgestellt. Ausführlich werden Schmelzofenkonstruktionen und Temperaturmeßmethoden verschiedener Art beschrieben und die Ofenwerte verschiedener Flugprogramme (ESA, NASA, Interkosmos) tabellarisch zusammengefaßt und verglichen. Verschiedene besondere Studien und Ergebnisse bisheriger Experimente enthält der vierte Teil („Case Studies and Results"). Das betrifft die Kristallisation von Eutektika, die Züchtung von Einkristallen usw.

Insgesamt wird ein guter informativer Überblick vermittelt, auch wenn die referierten Ergebnisse nicht ganz vollständig sind und bis auf das Sojus-Apollo-Experiment die vorhandene Literatur zum Interkosmosprogramm kaum in die Auswertung einbezogen wird. Die behandelten Probleme werden so dargestellt, daß sie auch für die irdische Experimentiertechnik und Modellierung von Interesse sind und oftmals Lehrbuchcharakter besitzen. Somit ist die Zusammenfassung nicht nur für Spezialisten der kosmischen Materialtechnologie, sondern auch für Materialforscher, Kristallzüchter, Metallurgen, Physiker, Chemiker, Biotechnologen und Gerätebauer in der irdischen Forschung von hohem Interesse.

P. Rudolph

Cryst. Res. Technol.	22	1987	7	877–882

N. S. Peev

Institute of Microelectronic, Sofia, Bulgaria

Morphology of SiC Epitaxial Layers Grown by Temperature Gradient Zone Melting (II)

The Dependence of the Structural Properties of the Epitaxial Layers on the Growth Conditions

In the present paper the relationships between the structural perfection of the layer, represented by the Tb average integral concentration (which is a criterion for the second phase quantity), the rocking curve half-width (a criterion for the general structural perfection of the layer) and the layer growth conditions, represented by the quantity, which is the most important characteristic of the growth process, namely the supersaturation are discussed. The considered model allows to conclude, that a optimal value of the supersaturation exists, at which the structural perfection is maximal.

В настоящей работе рассматривается зависимост структурного совершенство эпитоксиального слоя, представленного средней интегральной концентрацией тербия (которая является критерием о количестве включении второй фазы) и полушириной кривы качания (критерии об общем структурном совершенстве слоя) от условий его роста, представленные величиной, которая является найболее важной характеристикой процесса роста, а именно пересышение. Рассмотренная модель позволила сделать вывоб, что существует оптимальное значение пересыщения, при котором структурное совершенство максимально.

Introduction

On SiC epitaxial layers, grown by temperature gradient zone melting in vacuum (Peev et al.) the Tb average integral concentration, which is a criterion for the second phase quantity in the layers (the second phase quantity is higher when the average concentration is higher) and the rocking curve, which half-which is a criterion for the general structural perfection of the layer (the structural perfection is better if the rocking curve halfwidth is smaller) are measured (Peev et al.). These parameters were measured as a function of the following quantities: growth temperature, initial melt composition, and the substrate side. It is mentioned in the paper (Peev et al.) that some of these dependences are not monotonous and are characterized by extremas, which is a proof for their complex nature. There is no theory at present, which explains even qualitatively the dependence of the structural perfection of the layer on the growth conditions. There exist individual considerations, which explain some mechanisms for generation of dislocations (Kulish; Alexandrov; Tiller), which are the most common defects in semiconductor crystals. It is underlined in the above mentioned paper (Peev), that all dependences of the structural perfection of the SiC layer on the growth conditions are not connected with the participation of second phase inclusions: some extremes do not coincide, others are absent at all. This means that there exist other sources of dislocations (imperfections), which in some cases predominate over the second phase inclusions.

The discussion of this problem in our paper is qualitatively too and we hope that it will be useful for the physical explanation of the process.

Discussion

In the related literature four mechanisms of dislocations generation during the growth process are described: 1) dislocations, generated by the coalescence of two-dimensional nuclei; 2) dislocations, caused by the non-uniform incorporation of impurities in the growth step; 3) dislocations, caused by the presence of concentration gradients in the solid phase; 4) dislocations, taken over from the substrate (KULISH; ALEXANDROV; TILLER). In this paper only the first two mechanisms will be considered, because they depend directly on the layer growth conditions, and therefore they can be influenced by them.

It is shown (PEEV) that in the growth of SiC by temperature gradient zone melting the participation of screw dislocations is observed. The generation of dislocations by the coalescence of two-dimensional nuclei is also considered in the paper. The dislocation density generated in this way is given by the expression (KULISH; ALEXANDROV)

$$N_{\text{d1}} = 0.15 \left(\frac{J}{v_\text{n}}\right)^{2/3} \text{cm}^{-2}, \tag{1}$$

where J is the nucleus generation rate and the v_n the growth rate. The nucleus growth step is given by

$$d \cong 0.8 \left(\frac{v_\text{n}}{J}\right)^{1/3} \text{cm}. \tag{2}$$

From these equations it follows that the dependence of dislocation density on the nucleus step height is given by the expression

$$N_{\text{d1}} = 0.15 \left(\frac{0.8}{d}\right)^{2} \text{cm}^{2}. \tag{3}$$

The interferometer measurements show that along the substrate surface the step height changes in the range 600–1500 Å. When the step height is 1000 Å the dislocation density is in the order of 10^9 cm^{-2}. The value of the dislocation density is 10^4 to 10^5 cm^{-2}, calculated on the basis of the rocking curve half-width. Such a large difference between the calculated and the experimental values is explained by the circumstance that in our experiments the crystal is grown by screw dislocation mechanism, which nevertheless does not exclude the formation of nuclei.

It is more expedient to transform the equation (3) in the form

$$N_{\text{d1}} = A\sigma^{2/3} \quad \text{cm}^{-2}, \tag{4}$$

where the formula $d \sim \sigma^{-1/3}$ (CHERNOV; CHERNOV, LJUBOV) is used, A is a constant and σ is the relative supersaturation above the crystallization boundary. In this way the dislocation density, generated by this mechanism, depends on the supersaturation — the most important parameter of crystal growth.

The density of dislocations which are generated by the non-uniform impurity capture from the growth step is given by the expression (KULISH; TILLER)

$$N_{\text{d2}} = \frac{4k_0G\,|\Delta r|}{m^* r_\text{s}^2} \tag{5}$$

where k_0 is the segregation coefficient of the impurities, G the temperature gradient in the liquid phase, $|\Delta r| = |r_s - R_s|$, r_s and R_s are the atomic radii of the impurities and of the matrix substance, respectively. The impurity atom in this process is incorporated in the crystal matrix and replaces the atom of the basic material. m^* is the decrease of the phase transition temperature which corresponds to one atomic part of impurity.

If it is assumed that Tb is the only impurity in the liquid phase then with the aid of the data from PEEV et al. one can calculate the dislocation density, caused by this mechanism as 20 cm^{-2}. This value is much smaller than the observed dislocation density. Therefore the structural imperfection of the layers cannot be explained with this mechanism. It is more expedient to transform the equation (5) to

$$N_{d2} = \frac{4k_0G\,|\Delta r|}{m^*r_s^2}\,\frac{D}{C_0N_{eq}^2}\,\frac{C_0N_{eq}^2}{D} = B(T)\,\sigma^2 \quad \text{cm}^{-2} \tag{6}$$

where

$m^* = \frac{dT}{dC} = m^{-1}$ – the slope of the liquidus line

$\Delta N^2 = \frac{mGD}{C_0}$ – the supersaturation above the crystallization boundary (PEEV)

$\sigma = \frac{\Delta N}{N_{eq}}$ – the relative supersaturation above the crystallization boundary and N_{eq} is the equilibrium concentration (7)

Equation (6) is valid for the screw dislocation growth (PEEV).

The problem of the liquid phase capture from the crystallization boundary and of the formation of second phase inclusions is discussed by CHERNOV, TEMKIN and MELNIKOVA (1976, 1977). For SiC this discussion is made by PEEV (1979). There are three mechanisms to form second phase inclusions:

— the formation of second phase inclusions by non-equilibrium capture of impurity atoms from the layers,
— the formation of second phase inclusions by the capture of solid particles, which are in the liquid phase just in front of the crystallization boundary,
— the formation of second phase inclusions by capturing parts of the liquid phase.

If the deviation from the state of equilibrium is too large, then the first mechanism takes place. In this case the value of the coefficient of segregation of the impurities may considerably differ from its equilibrium value and therefore the impuritie concentration in the solid phase may significantly exceed its equilibrium value. Then by solid phase diffusion the second phase inclusions are formed. In the SiC liquid phase epitaxy by temperature gradient zone melting the growth rate and the supersaturation at the crystallization boundary are small (PEEV 1986; 1979), therefore the deviation from the state of equilibrium is small too, and the formation of second phase inclusions by this mechanism is impossible.

The solid phase particles in the melt may be formed by chemical reactions in the liquid phase. If they are close to the crystallization boundary in some circumstances these particles could be captured by the crystallization boundary. In order to determine the following factors one has to be taken into account (CHERNOV, TEMKIN, MELNIKOVA 1976, 1977): the hydrodynamical forces with which the crystallizing component in its diffusion to the crystallization boundary presses the particles to the growth interface; the decrease of the supersaturation in the region between the solid particles and the crystallization boundary; the Gibbs-Thompson effect a.o. The analysis of the obtained equation with the well known characteristics of the Tb–Si–SiC system (PEEV 1979; PEEV et al. 1986, 1987) have shown that by using a melt with initial composition 100 at% Tb a microparticle with radius R will be captured if the growth rate is higher then the critical value V_c, given by

$$V_c = \frac{7 \cdot 10^{-17}}{R^2} \ (\text{cm} \cdot \text{s}^{-1}) \,. \tag{8}$$

For initial melt composition 85 at.% Si–15 at.% Tb the critical value V_c is equal to

$$V_c = \frac{4 \cdot 10^{-16}}{R^2} \ (\text{cm} \cdot \text{s}^{-1}) \,. \tag{9}$$

At $R = 1000$ Å the equation (8) gives $V_c = 0.5\ \mu\text{m} \cdot \text{min}^{-1}$ and the equation (9) $-V_c = 2.5\ \mu\text{m} \cdot \text{min}^{-1}$. The last two equations show that the second phase inclusions quantity increases with the increasing of the growth rate: the larger growth rate value leads to the capture of smaller particles. Therefore one may conclude that the quantity of the captured particles is proportional to the growth rate and, respectively, to the supersaturation ($V_c \sim \sigma^2$ in the growth by screw dislocations mechanism)

$$N' = C\sigma^s \,. \tag{10}$$

The parameter s may be larger, equal to or smaller than unity.

The problem of the liquid phase capture from the crystallization boundary is discussed qualitatively by CHERNOV and by CHERNOV and LJUBOV where a mechanism for the formation of second phase inclusions is described. The change of the growth step height (the step of the two-dimensional nuclei or of the screw dislocations) with the change of the supersaturation, which is given by the expression $d \sim \sigma^{-1/3}$ creates a condition for the capture of the liquid phase by the crystallization boundary. Having in mind this dependence one can make the following conclusions: when the supersaturation increases the growth step height decreases; for low supersaturation (see PEEV 1979; PEEV et al. 1986, 1987; it is shown that the growth of SiC from Si–Tb solutions by temperature gradient zone melting takes place at low supersaturation 0.01–0.001%) the slope of this dependence is too steep, and a slight change of supersaturation leads to a significant change of the growth step height. For high supersaturations the growth step height almost does not change with the change of supersaturation. If the growth step, spreading on the surface of the crystallization boundary, reaches a region with higher supersaturation then it disintegrates into two steps with lower growth step height (Fig. 1). In the liquid phase epitaxy this disintegration takes place primarily through the formation of an overhanging layer (CHERNOV; CHERNOV, LJUBOV). With the formation of the overhanging layer the part of EGC (Fig. 1) lags behind the step because it is unsufficiently fed with a crystallizing component, and after the contact of this growth step with another one this part will remain in the solid

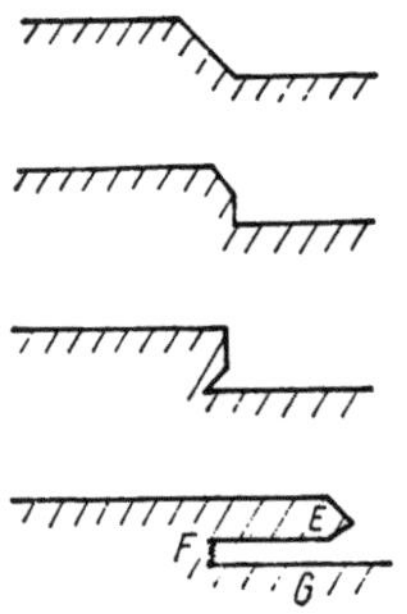

Fig. 1. Liquid phase capture from the crystallization boundary and formation of second phase inclusion

phase. The measurement of the growth step height along the interface boundary indicated that the growth step height in a given layer changed significantly, and it was in the region of 600–2000 Å. Therefore the formation of second phase inclusions with dimensions 400–1000 Å is quite possible.

During the cooling after the growth the deviation from the state of equilibrium in the liquid phase is too large, and the supersaturation above the crystallization boundary has been considerably increased. Because of the dependence $d \sim \sigma^{-1/3}$ the measured value of the growth step is lower and its real value is larger. If one takes into account this circumstance one may conclude that in the region of the slight supersaturation the quantity of the liquid phase inclusions in the epitaxial layer will be increased: the quantity of the liquid phase inclusions is inversely proportional to the supersaturation:

$$N'' = D\sigma^{-p} . \tag{11}$$

During the cooling period after the stopping of growth between the second phase inclusions and the crystal matrix thermal stresses appear. These stresses lead to the deformation of the crystal and to the generation of dislocations. This problem is discussed in detail by GEGUSIN and DSUBA (1973, 1974), by GEGUSIN and CRIVOGLAS, and by CACHANOV.

The higher quantity of second phase inclusions in the epitaxial layer causes a higher dislocations density. Therefore the following equations can be written:

$$N_{d3} = E\sigma^m \quad \text{cm}^{-2} \tag{12}$$

from equation (10) and

$$N_{d4} = F\sigma^{-n} \quad \text{cm}^{-2} \tag{13}$$

from equation (11).

The parameters E and F depend on the temperature, the melt composition, the mechanical parameters of the layer and so on. The equations (4), (6), (12) and (13) disclose the dependence of the dislocations density on the supersaturation. The sum of these four equations gives the total dislocations density of the layer

$$\begin{aligned} N_d &= N_{d1} + N_{d2} + N_{d3} + N_{d4} = \\ &= A\sigma^{2/3} + B(T)\,\sigma^2 + E\sigma^m + F\sigma^{-n} . \end{aligned} \tag{14}$$

The analysis of the equation (14) shows that at high supersaturation the defects in the layer are caused by the first three mechanisms (equations (4), (6), (12)), whereas at low supersaturation the fourth mechanism (eq. (13)) predominates. This means that the dependence of the dislocations density on the supersaturation above the crystallization boundary must be minimum with the value $\sigma = \sigma^*$ (Fig. 2).

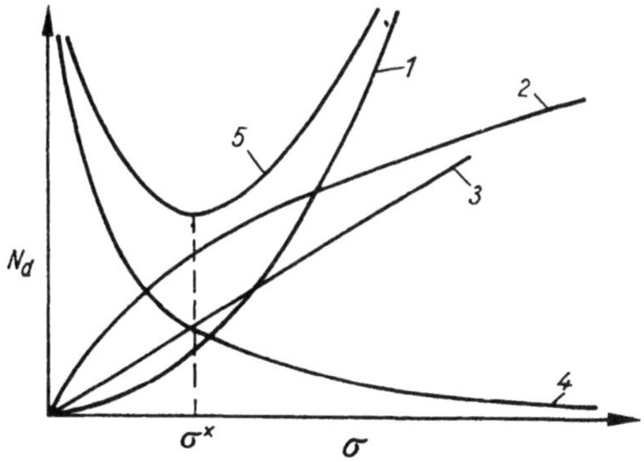

Fig. 2. Structural perfection of the epitaxial layer, given by the dislocation density versus the supersaturation above the crystallization boundary: (1) N_{d2} given by Eq. 6; (2) N_{d1} given by Eq. 4; (3) N_{d3} given by Eq. 12 – it is assumed $m = 1$; (4) N_{d4} given by Eq. 13; (5) the general structural perfection, given by Eq. 14

It must be noted that equations (6) and (14) are related to the liquid phase epitaxy by temperature gradient zone melting, as the equation (7) concerns this growth method. By using other LPE methods the same minimum may be expected, but a different value of σ^*.

Summary

In order to check the validity of equation (14) and the curve from Figure 2 it is necessary to change only the supersaturation above the crystallization boundary (for example by changing the temperature gradient), whereas the growth temperature and the initial melt composition remain constant. In this way it is possible to determine the optimal supersaturation σ^* for given growth conditions.

In the experiments described by PEEV et al. (1986, 1987) the initial melt composition and the growth temperature have been changed as well; therefore they can be found to the left or to the right of the minimum or at the minimum itself. Because of that the origination of extremas is quite possible.

References

ALEXANDROV, L. N.: Kinetica crystall. i struct. epitaxialnich sl. dvuchcomponentnich poluprovodnicov, in: Rost crystallov **10**, 241 (1968)

CACHANOV, L. M.: Ocnovi theorij plastichnosti, Moscow 1969, p. 420

CHERNOV, A. A.: Uspechi phys. nauk **73** (2) 277 (1961)

CHERNOV, A. A., LJUBOV, B. Ja.: in: Rost crystallov **5**, 11 (1965)

CHERNOV, A. A., TEMKIN, D. E., MELNIKOVA, A. M.: Crystallographiya **21**, (4) 652 (1976)

CHERNOV, A. A., TEMKIN, D. E.: Crystallographiya **22**, (5) 933 (1977)

CHERNOV, A. A., TEMKIN, D. E., MELNIKOVA, A. M.: Crystallographiya **22**, (6) 1152 (1977)

GEGUSIN, Ja. E., DSUBA, A. S.: Crystallographiya **19**, (5) 1030 (1974)

GEGUSIN, Ja. E., DSUBA, A. S.: Crystallographiya **18**, (4) 813 (1973)

GEGUSIN, Ja. E., CRIVOGLAS, M. A.: Dvishenie macroscopich. Vkljuchenij v tverdich telach, Moscow 1971, p. 344

KULISH, U. M.: Rost i electrophys. svoistva plenok poluprovodn. (shidkostnaja epitaxija), Elista 1976, p. 205

PEEV, N. S., TAIROV, YU. M., SMIRNOVA, N. A., KALNIN, A. A.: parts 1, 2, and 3 in: Cryst. Res. Technol. **21**, 1503, 1509 (1986) and **22**, 59 (1987)

PEEV, N. S.: Cryst. Res. Technol. **21** (1) 319 (1986)

PEEV, N. S.: Dissertation, Electrotechnical Inst., Leningrad 1979

TEMKIN, D. E., CHERNOV, A. A., MELNIKOVA, A. M.: Crystallographiya **22**, (1) 27 (1977)

TILLER, W.: J. Appl. Physics **29**, 611 (1958)

(Received November 4, 1986)

Author's address:

Dr. N. S. PEEV

Institute of Microelectronic
Boulvd. Lenin 7 km
Sofia 1784
Bulgaria

Cryst. Res. Technol.	22	1987	7	883–892

H. KÜHNE

Akademie der Wissenschaften der DDR
Institut für Halbleiterphysik, Frankfurt (Oder)

The Current Understanding of Epitaxial CVD Silicon Layer Doping in the Light of Modelling and Theory Development (IV)

The Problem of More-than-one Kind of Dopant Species

In Part IV of the present treatise epitaxial silicon layer doping governed by the rate-limiting action of more than one kind of dopant species in the gas is investigated on the theoretical basis which has been derived in the previously published parts. It is shown that both the general reaction scheme of forward and backward reaction steps and the general differential equation statement of counterbalancing the fluxes of those reaction steps and the dopant incorporation flux are suitable for solving the problem under consideration. Only two of the four incorporation-limiting reaction mechanisms, being discussed in the more recent literature, fulfill the requirements needed to agree with experimental findings on the one hand and with the limiting case situation of equilibrium incorporation on the other. Arsenic has been selected as the demonstrating dopant element.

Im vierten Teil der vorliegenden Arbeit wird die Dotierung epitaktischer Siliciumschichten für den Fall untersucht, daß mehr als nur eine Art von Dotandenspezies im Gas bestimmend am Einbaugeschehen teilnimmt. Hierfür wird die in den vorangegangenen Teilen entwickelte theoretische Grundlage verwendet. Sowohl das allgemeine Reaktionsschema der Hin- und Rückreaktionen als auch der allgemeine Differentialgleichungsansatz zur Bilanz der Reaktionsströme und des Dotierungseinbaustromes sind für die Lösung des betrachteten Problems erforderlich. Nur zwei der vier in der gegenwärtigen Literatur vorgeschlagenen, den Einbau begrenzenden Reaktionsmechanismen erfüllen die Forderungen, die zu Einbaugesetzen führen, welche einerseits mit den experimentellen Befunden und andererseits mit dem Grenzfall des Gleichgewichtseinbaus übereinstimmen.

Introduction

In the previously published Parts I–III of this report the empirical fundamentals of the epitaxial CVD silicon doping process were combined with relevant parts of the basis knowledge of sciences in such a way that a general mechanism of dopant incorporation could be derived and a general differential equation expression could be formulated (12/II)[1])

$$\frac{dI_{\text{dope}}}{dt}\cdot\frac{dt}{dx}=\frac{1}{v_{\text{epi}}}\cdot[I_1-(I_2+I_{\text{dope}})]\,. \qquad (12/\text{II})$$

[1]) Roman numbers refer to the previously published parts of the treatise.

Under steady state conditions of doping the forward (I_1) and the backward directed reaction (I_2) of one of the reaction steps in the sequence throughout the overall reaction is balanced by the dopant incorporation flux (I_{dope}). The doping process as a whole covers the transport of the dopant source substance in the gas, its decomposition by heat, the interdependence of those decomposition products with the surface of the growing layer, the formation and fixation of dopant atoms on the surface, and their incorporation into bulk lattice sites. The differential equation (12/II) can be solved for each particular rate limiting step formulated as has been shown in Part III of this report. The hypothetical incorporation laws, derived in this way from the most important conceptions discussed in more recent literature, can be applied to practical problems after empirical constants having been substituted for the theoretical quantities. Thus the different hypothetical laws are changed into one and the same unique model equation (1).

$$(N_{dope})_{total} = \frac{A \cdot p^0_{XH_3}}{(p_{total} + B \cdot v_{epi})} \cdot \left[1 - \exp \frac{a}{v_{epi}} \cdot (p_{total} + B \cdot v_{epi}) \cdot \Delta x\right], \tag{1}$$

where:

$$a(p_{total} + B \cdot v_{epi}) \equiv \frac{v_{epi}}{L} = a' + b' \cdot v_{epi}; \qquad p_{total} \equiv p^0_{H_2} \approx p^{eq}_{H_2}.$$

The above theory-aided model equation describes dopant incorporation not only for steady state ($\Delta x \gg L; t \to \infty$) but also for non-steady state conditions ($0 < \Delta x < 10L$). It possesses a more universal importance compared to the fairly good equation derived formerly by REIF et al. by means of empirical methods, and by WONG et al. as well as by KÜHNE (1) by means of analytical methods, for the total pressure ($p_{total} \approx p^0_{H_2} \approx \approx p^{eq}_{H_2}$) has taken into account now. The total pressure, an important process controlling parameter, is introduced into equ. (1) by substituting for the equilibrium pressure of hydrogen, which, in case that excessive amounts of hydrogen are used as the carrier gas, represents the hydrogen input pressure in the same way as the total pressure.

In Part II of the treatise attention has been paid to the circumstance that possibly more than one species take part in the dopant incorporation reaction. But the different hypothetical laws derived in Part III are related to the situation that only one kind of particles, containing dopant atoms, controls the incorporation reaction.

The present paper is concerned with the extension of the dopant incorporation equation to conditions characterized by the action of more than one kind of dopant species. Furthermore, consequencies are to be investigated which might arise from such an extended law with respect to the non-distinguishably equally good, but different mechanisms having been treated in Part III as those by which the doping process possibly should be determined in reality. Yet, as a matter of fact, doping can be controlled only by one of them.

A brief historical review

The fact that dopant source substances as phosphine, arsine, and diborane are decomposed by heat already below the range of typical temperatures of silicon epitaxy had been used by RAI-CHOUDHURY et al. to explain differences of the incorporation behaviour in the ranges of lower and higher input pressures of the source substance.

$$(N_{dope})_{total} = \text{const} \cdot (p^0_{XH_3})^{1/i}, \tag{1/I}$$

where:

$i = 1$ and 2.

In the higher range of input pressures of the dopant source biatomic element particles were assumed to be dominant in the decomposition equilibrium, while in the range of lower input pressures atomic element particles should dominate

$$XH_3 \rightarrow X \rightleftharpoons 1/2\,X_2 \rightleftharpoons 1/4\,X_4\,. \tag{2a}$$

Calculations of equilibrium compositions of the P–H system, carried out by BLOEM et al., showed that P–H containing species play a dominant role at least in the lower range of PH_3-input pressures

$$XH_3 \rightleftharpoons XH_2 \rightleftharpoons XH \rightleftharpoons X \rightleftharpoons 1/2\,X_2 \rightleftharpoons 1/4\,X_4\,. \tag{2b}$$

It suggests itself that thermodynamic equilibrium possibly should include the step of dopant incorporation too. Therefore, the application of the Mass Action Law to the problem was a logical consequence of the above cited authors

$$(N_{\text{dope}})_{\text{epi}} = K_X \cdot p_X^{\text{eq}}\big|_{p^0_{XH_3} < (p^0_{XH_3})_{\text{crit.}}}\,, \tag{3a}$$

$$(N_{\text{dope}})_{\text{epi}} = K_{X_2} \cdot \sqrt{p_{X_2}^{\text{eq}}}\,\big|_{p^0_{XH_3} > (p^0_{XH_3})_{\text{crit.}}}\,. \tag{3b}$$

The question arises why X- and X_2-species should play an extraordinary role compared to the other components in the system. As shown in the Figures 1, 2, and 3 the main components are the hydrids PH_3 and PH_2, the equilibrium pressures of which are nearly equal, though the concentration of the PH_2-species grows at the expence of that of the PH_3-species when the temperature is increased. In the lower range of

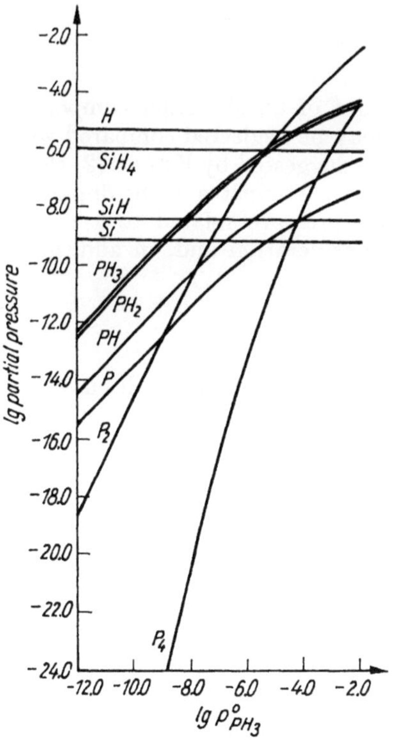

Fig. 1. Calculated equilibrium compositions of the P–H system as a function of the input pressure of phosphine, $p_{\text{total}} = 1$ atm., $T = 1400$ K (BLOEM et al.)

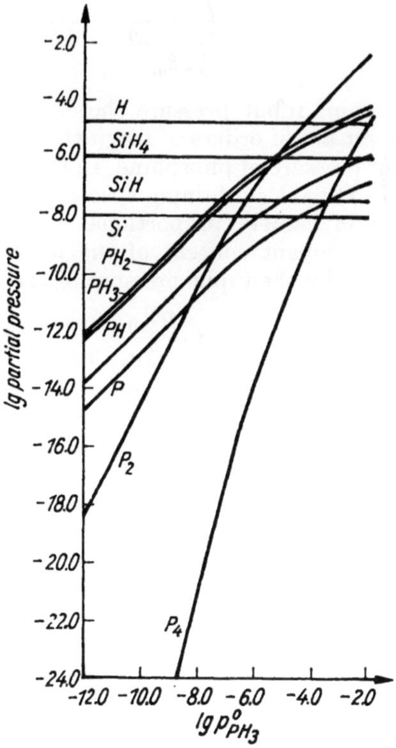

Fig. 2. The same dependencies as shown in Figure 1, but $T = 1500$ K (BLOEM et al.)

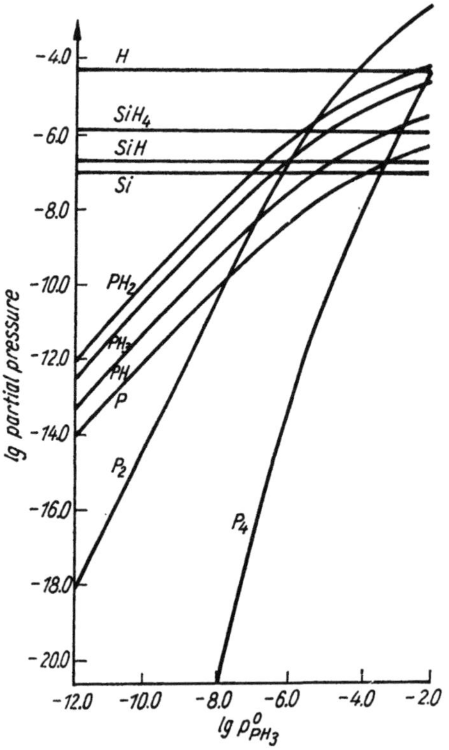

Fig. 3. The same dependencies as shown in Figure 1, but $T = 1600$ K (BLOEM et al.)

phosphine input pressure the equilibrium pressures of all the other components of interest are of orders of magnitude smaller and may be neglected compared with the input pressure of phosphine. In that context it was suggested by KÜHNE (2) that each of the dopant containing species existing in the gas participates in the dopant incorporation namely proportional to its equilibrium pressure related to a quasi-one-atomic dopant content of the particles under consideration and, in addition, normalized by the input pressure of the dopant source:

$$y_{X_iH_j} = i \cdot \frac{p^{eq}_{X_iH_j}}{p^0_{XH_3}}. \tag{4}$$

Such an assumption is compatible with an equilibrium situation as discussed above if it is supposed by tacit agreement that the equilibrium in the gas is connected with the equilibrium of incorporation by means of a kinetic mechanism of dopant transport. In such a way the equilibrium concept proved to be useful in the cases of phosphorus (KÜHNE, 3) and arsenic (KÜHNE et al.) when theoretical dopant incorporation enthalpies were compared with experimental findings. Dopant incorporation has been explained by applying the equilibrium concept not only for atmospheric pressure but also for reduced pressure conditions (KÜHNE, 4).

Contrary to the ideas given above, many authors had considered dopant incorporation analogous to layer growth being kinetically governed by gas transport or surface reactions. REIF et al. emphasized the point of view that dopant incorporation must include a transport mechanism. Therefore, it should be interpreted by a mainly kinetical process mechanism. First they supposed a rather passive behaviour of dopants which, fixed on the surface by adsorption, should be burried by deposited silicon atoms. The dopant transport was thought to be realized by the dopant adsorption flux. From their investigations the influence of layer growth on dopant incorporation was confirmed and, additionally, a theoretical explanation of the experimental findings given. Furthermore they extended their investigations to non-steady state dopant incorporation behaviour, treading the dopant process as a black-box and applying methods of cybernetics to the problem.

Coworking with DUTTON, REIF derived an analytical basis to the incorporation model discussed. The results enabled them to implement a computer programm in that Fick's Second Law had been included. It is succesfully applied to practical use in simulating dopant profiles. Recently in coworking with WONG, REIF extended the analytical treatment of the non-steady state doping process to lateral autodoping in order to give a theoretical explanation of the phenomenon.

Previous to this KÜHNE (5) had pointed out that a theoretical model of the doping process, being based on reaction kinetics, must result from taking into account not only such reaction steps which contribute to incorporation but also those which are opposed to them. He also showed, taking phosphorus as an example, that the layer growth influence on doping depends on temperature in such a way that in the range of high deposition temperatures, which corresponds to $SiCl_4$-epitaxy, it is negligibly small.

Decomposition of the dopant source substance

As mentioned above phosphine decomposes in the range of lower input pressures mainly forming PH_3- and PH_2-species. They participate to nearly the same degree in epitaxial layer doping on the one hand and, in a practical sense, totally balance the input pressure of phosphine on the other

$$p^0_{PH_3} \approx p^{eq}_{PH_3} + p^{eq}_{PH_2} . \tag{5}$$

Arsine is decomposed by heat in a hydrogen atmosphere under conditions of reduced pressure forming not only AsH_2-species but also As-species in a comparable degree (Fig. 4)

$$p^0_{AsH_3} \approx p^{eq}_{AsH_2} + p^{eq}_{As} . \tag{6}$$

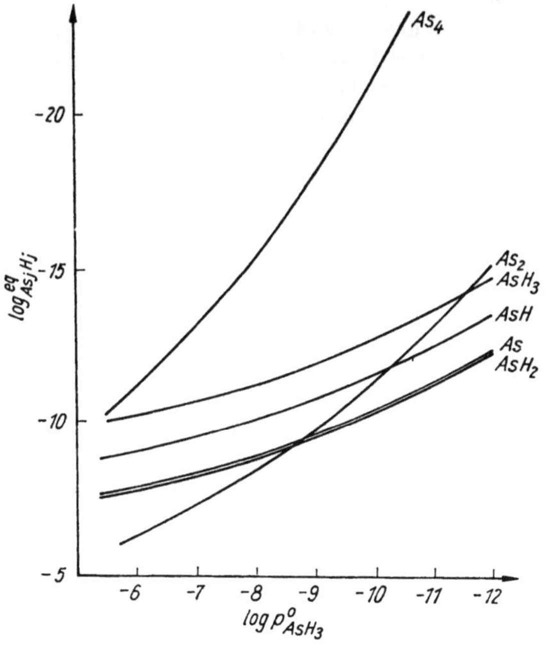

Fig. 4. Calculated equilibrium compositions of the As–H system as a function of the input pressure of arsine, $p_{total} = 0.01$ atm, $T = 1300$ K (KOKOVIN et al.)

The decomposition of diborane mainly yields BH_3- and BH_2-species in nearly the same concentrations (cp. Fig. 2/II) as has been calculated by HINTZEN and coworkers. Contrary to phosphorus and arsenic boron also forms chlorine and/or oxygen containing products if those elements are present in the gas. Nevertheless BH_3- and BH_2-species are the main components of the pure hydrids in the system, and they are considered to be the incorporation-active species. MORGENSTERN et al. lately published results of thermodynamic calculations from which only BH_3 should be present as the main component (cp. Fig. 3/II). But in the context discussed here the results of HINTZEN et al. are of more interest

$$2 \cdot p^0_{B_2H_6} \approx p^{eq}_{BH_3} + p^{eq}_{BH_2} \,. \tag{7}$$

The general dopant incorporation law

In the following the four different incorporation-limiting mechanisms having been discussed in Part III of the treatise will be considered on condition that two different kinds of dopant particles participate in the doping process one of them being the dopant element itself. This latter assumption is accepted in the case of arsenic under reduced total pressure. For our purpose the differential equation statement (12/II) has to be used following the scheme that has been published in Part III.

As to the "simple" conception of dopant incorporation, it is characterized by the transfer of arsenic atoms from the gas into the silicon layer. This concept contradicts the basic assumption that at least two different species partake in dopant incorporation. It does not allow any As-containing particle, except As-species itself, to be involved in the dopant incorporation reaction. Therefore, the simple conception is not applicable in the special case considered.

The "extended" conception allows all the kinds of dopant atoms containing species to take part in forming arsenic atoms by chemical reactions within the gas. Though, strictly spoken, only the main components need to be considered. The arsenic atoms in the gas are steadely consumed by incorporation in the silicon layer. Yet not the latter but the former reaction step is regarded as the overall rate limiting step. This extended conception is also not applicable to the problem of interest, for the incorporation step is characterized by an equilibrium-like mechanism. So, if arsenic atoms would dominate in the gas, incorporation should proceed as an equilibrium process as shown below:

$$(As)_g \rightleftharpoons As_{Si} \,, \tag{8a}$$

$$(N_{As_{Si}})_{T_{epi}} = K_{As} \cdot p^{eq}_{As} \approx K_{As} \cdot p^0_{AsH_3} \,. \tag{8b}$$

Now assume As- and AsH_2-species present in the gas with nearly equal concentrations, the incorporation flux consisting of As-species would have to be feeded to a certain portion by a spontaneous decomposition of the source substance to As-particles on one hand and by the kinetically controlled chemical reaction of AsH_2-species to As-particles on the other. Combining both partial fluxes leads, in a formal way, from the equs. (6) and (8) to equ. (9/III) by which neither the experimental findings (cp. equ. 5/I) nor theoretical results (cp. equ. 6/II) are expressed correctly

$$(N_{As_{Si}})_{T_{epi}} = \frac{K_{As} \cdot K_{AsH_2} \cdot p^0_{AsH_3}}{\left(p_{total} + \dfrac{K_{As}}{F_1 \cdot k} \cdot v_{epi} + K_{AsH_2}\right)} \,. \tag{9/III}$$

In the contrast to the both conceptions discussed above, the "consequent" conception allows an equal and independent treatment of both kinds of species following the course of the general incorporation reaction scheme derived in Part II of the present work

$$(\mathrm{As})_g \rightarrow \mathrm{As}_{\mathrm{Si}}\,, \qquad I_1 = k_1 \cdot p_{\mathrm{As}}^{\mathrm{eq}}\,, \tag{9a}$$

$$\mathrm{As}_{\mathrm{Si}} \rightarrow (\mathrm{As})_g\,, \qquad I_2 = k_2 \cdot (N_{\mathrm{As_{Si}}})_{T_{\mathrm{epi}}}\,, \tag{9b}$$

$$\mathrm{AsH}_2 \rightarrow \mathrm{As}_{\mathrm{Si}} + \mathrm{H}_2\,, \qquad I_3 = k_3 \cdot p_{\mathrm{AsH_2}}^{\mathrm{eq}}\,, \tag{9c}$$

$$\mathrm{H}_2 + \mathrm{As}_{\mathrm{Si}} \rightarrow \mathrm{AsH}_2\,, \qquad I_4 = k_4 \cdot p_{\mathrm{H_2}}^{\mathrm{eq}} \cdot (N_{\mathrm{As_{Si}}})_{T_{\mathrm{epi}}}\,, \tag{9d}$$

$$I_{\mathrm{dope}} = v_{\mathrm{epi}} \cdot (N_{\mathrm{As_{Si}}})_{\mathrm{total}} = \frac{v_{\mathrm{epi}}}{F_1} \cdot (N_{\mathrm{As_{Si}}})_{T_{\mathrm{epi}}}\,. \tag{9e}$$

The equations (9a)—(9e) introduced in the general differential equation (12/II) yield a still more general version of the latter

$$\frac{dN_{\mathrm{As_{Si}}}}{dt} \cdot \frac{dt}{dx} = \frac{1}{v_{\mathrm{epi}}} \cdot \{(I_1 + I_3) - [(I_2 + I_4) + I_{\mathrm{dope}}]\}\,. \tag{10}$$

Equ. (10) is solved by integration within the borders $t = 0 \equiv N = 0$ and $t = \infty \equiv \equiv N = N_\infty$ leading to:

$$(N_{\mathrm{As_{Si}}})_{\mathrm{total}} = \frac{k_1 \cdot \left(p_{\mathrm{As}}^{\mathrm{eq}} + \frac{k_3}{k_1} \cdot p_{\mathrm{AsH_2}}^{\mathrm{eq}}\right)}{F_1 \cdot k_2 \cdot \left(1 + \frac{k_4}{k_2} \cdot p_{\mathrm{H_2}}^{\mathrm{eq}} + \frac{v_{\mathrm{epi}}}{F_1 \cdot k_2}\right)} \cdot \left[1 - \exp -\frac{k_2}{v_{\mathrm{epi}}} \cdot \left(1 + \frac{k_4}{k_2} \cdot p_{\mathrm{H_2}}^{\mathrm{eq}} + \frac{v_{\mathrm{epi}}}{F_1 \cdot k_2}\right) \cdot \Delta x\right]. \tag{11}$$

The somewhat complicated looking equ. (11) might be simplified by the following both reflections. The reaction rate constants given by the equs. (9a)—(9d) may be used to define equilibrium constants:

$$\frac{k_1}{k_2} = K_{\mathrm{As}}; \tag{12a}$$

$$\frac{k_3}{k_4} = K_{\mathrm{AsH_2}}; \tag{12b}$$

$$\frac{k_4}{k_2} = \frac{K_{\mathrm{As}}}{K_{\mathrm{AsH_2}}}. \tag{12c}$$

Furthermore it is allowed to conclude from equ. (4) the statement given by equ. (13) since the former equation is based on the presumption that all the chemical reactions involved are not limited by kinetical effects

$$k_1 \approx k_3 \ldots; \tag{13a}$$

$$\frac{k_3}{k_1} = 1\,. \tag{13b}$$

Restricting equ. (11) to steady state incorporation, introducing the equs. (12) and (13) and taking into account equ. (6) leads to the following expression:

$$F_1 \cdot (N_{\mathrm{As_{Si}}})_{\mathrm{total}} = (N_{\mathrm{As_{Si}}})_{T_{\mathrm{epi}}} = \frac{K_{\mathrm{As}} \cdot p_{\mathrm{AsH_2}}^0}{\left(1 + \frac{K_{\mathrm{As}}}{K_{\mathrm{AsH_2}}} \cdot p_{\mathrm{total}} + \frac{v_{\mathrm{epi}}}{F_1 \cdot k_2}\right)}\,, \tag{11a}$$

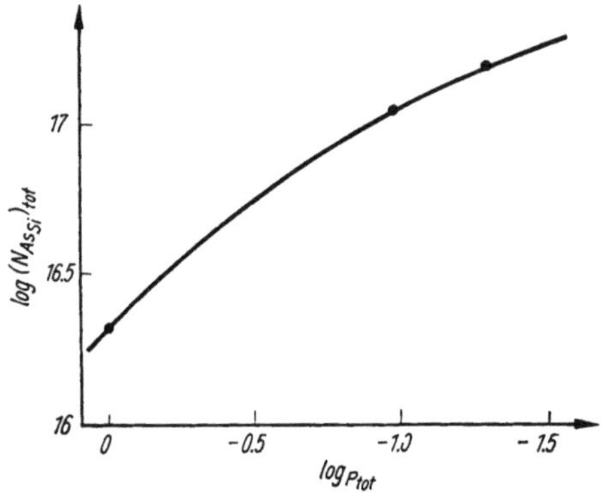

Fig. 5. Experimental dependence of arsenic incorporation in a hydrogen atmosphere on total pressure compared with theory (———). The partial pressure of arsine was held at a constant level. $T = 1353$ K (HERRING; KOKOVIN et al.)

where:

$$p_{\text{total}} \approx p_{H_2} \approx p_{H_2}^{\text{eq}} .$$

Equation (11a) describes the incorporation behaviour of arsenic in a hydrogen atmosphere under reduced pressure conditions. It is correct because its reduction to the limiting case of equilibrium incorporation, for what case the v_{epi}-term has to be neglected, yields the correct equ. (6/II). The latter is the result of the theoretical treatment of the problem on the basis of an equilibrium situation (cp. Part II). It also agrees with experimental findings (cp. Fig. 5).

Finally, the "improved" conception confines the kinetically controlled reaction step to the formation of an adsorbed layer of dopants. Adsorption and desorption fluxes may be described by linear relationships since the silicon surface is scarcely covered with adsorbed particles (Henry region of adsorption), what has been proved true by CLAASEN and coworkers. In the scheme followed here adsorbed dopant atoms have to be considered as the incorporation-active species. In agreement with equ. (4) the concentration of adsorbed arsenic atoms is maintained by the action of all the different dopant species in the gas. The arsenic incorporation into the growing silicon layer lastly occurs by means fo a solution equilibrium. Contrary to the situation given by the "extended" conception, in the present case arsenic atoms in the gas do not directly take part in the special incorporation step.

The reaction scheme now is as follows:

$$* + (\text{As})_g \rightarrow \text{As}^* , \qquad I_1 = k_1 \cdot (N_0 - N_*) \cdot p_{\text{As}}^{\text{eq}} \approx k_1 \cdot N_0 \cdot p_{\text{As}}^{\text{eq}}, \quad (14\text{a})$$

$$\text{As}^* \rightarrow (\text{As})_g + * , \qquad I_2 = k_2 \cdot N_{\text{As}^*} , \quad (14\text{b})$$

$$* + \text{AsH}_2 \rightarrow \text{As}^* + \text{H}_2 , \qquad I_3 = k_3 \cdot (N_0 - N_*) \cdot p_{\text{AsH}_2}^{\text{eq}} \approx k_3 \cdot N_0 \cdot p_{\text{AsH}_2}^{\text{eq}}, \quad (14\text{c})$$

$$\text{H}_2 + \text{As}^* \rightarrow \text{AsH}_2 + * , \qquad I_4 = k_4 \cdot p_{\text{H}_2}^{\text{eq}} \cdot N_{\text{As}^*} , \quad (14\text{d})$$

$$\text{As}^* \rightleftharpoons \text{As}_{\text{Si}} + * , \qquad K_{\text{inc}} = \frac{N_0 \cdot (N_{\text{As}_{\text{Si}}})_{T_{\text{epi}}}}{N_{\text{As}^*}} , \quad (14\text{e})$$

$$I_{\text{dope}} = v_{\text{epi}} \cdot (N_{\text{As}})_{\text{total}} = \frac{v_{\text{epi}}}{F_1} \cdot (N_{\text{As}_{\text{Si}}})_{T_{\text{epi}}} = \frac{v_{\text{epi}} \cdot K_{\text{inc}}}{F_1 \cdot N_0} \cdot N_{\text{As}^*} . \quad (14\text{f})$$

Another useful dopant incorporation law can be derived from the above set of equations:

$$(N_{\mathrm{As_{Si}}})_{\mathrm{total}} = \frac{K_{\mathrm{inc}} \cdot (K_{\mathrm{As}})_{\mathrm{ad}} \cdot p^{0}_{\mathrm{AsH_3}}}{F_1 \cdot \left[1 + \frac{(K_{\mathrm{As}})_{\mathrm{ad}}}{(K_{\mathrm{AsH_2}})_{\mathrm{ad}}} \cdot p^{\mathrm{eq}}_{\mathrm{H_2}} + \frac{K_{\mathrm{inc}}}{F_1 \cdot N_0 \cdot k_2} \cdot v_{\mathrm{epi}}\right]} \cdot \left\{1 - \exp - \frac{k_2}{v_{\mathrm{epi}}} \cdot \left[1 + \frac{(K_{\mathrm{As}})_{\mathrm{ad}}}{(K_{\mathrm{AsH_2}})_{\mathrm{ad}}} \cdot p^{\mathrm{eq}}_{\mathrm{H_2}} + \frac{K_{\mathrm{inc}}}{F_1 \cdot N_0 \cdot k_2} \cdot v_{\mathrm{epi}}\right] \cdot \Delta x\right\}. \tag{15}$$

Equ. (15) is as useful as equ. (11a) what may be demonstrated by substituting empirical constants for the theoretical ones

$$(N_{\mathrm{As_{Si}}})_{\mathrm{total}} = \frac{A \cdot p^{0}_{\mathrm{AsH_3}}}{(1 + B \cdot v_{\mathrm{epi}} + C \cdot p_{\mathrm{total}})} \cdot \left[1 - \exp - \frac{\Delta x}{L}\right] \tag{16}$$

where:

$$\frac{v_{\mathrm{epi}}}{L} = a \cdot (1 + B \cdot v_{\mathrm{epi}} + C \cdot p_{\mathrm{total}}); \qquad a = k_2; \qquad p_{\mathrm{total}} \approx p^{0}_{\mathrm{H_2}} \approx p^{\mathrm{eq}}_{\mathrm{H_2}}.$$

Equ. (16) might be compared to equ. (1). Both equations represent special solutions of the general differential equation statement for different compositions of the decomposition equilibrium of the dopant source substance.

Conclusions

Applying the incorporation scheme derived in Part II of the present work we are able to describe epitaxial silicon layer doping not only for incorporation processes controlled by no more than one, but also for processes governed by more than one kind of dopant species in the gas. Taking arsenic as an example it has been shown that only those incorporation mechanims may be accepted as possibly correct which allows the element atom species to react independently of and parallel to other kinds of dopant containing species. This requirement is fulfilled by transferring element atoms from the decomposition products of the dopant source substance in the gas into the silicon bulk in a direct way ("consequent" conception). It can also be fulfilled by any mechanism which agrees with the three-step-scheme given below:

— Decomposition of the dopant source substance and formation of the thermodynamic equilibrium composition.

— Kinetically controlled transport of the species, formed by the decomposition reaction, from the gas to the silicon surface where dopant atoms are formed. Part of them are fixed in a state not identical with that of the dopant atoms in the gas.

— Incorporation of dopant atoms by transferring them from that intermediate state to that of having occupied bulk lattice sites by means of an equilibrium-like process step.

The intermediate state, imagined here, must not necessarily be characterized by adsorption on the surface, but relevant terms and pictures are commonly used in literature what makes that conception superior to others.

The three-step-scheme should advantageously be used if, as suggested by Kühne (6), autodoping is interpreted to be the opposite process to intentional layer doping. Assuming no dopants present in the gas the surface dopant concentration of, for instance, burried layers determines the dopant concentration of such adsorption layer. The latter causes a dopant flux into the gas which has to be considered to be the origin of unintentional doping effects outside of burried layers.

Consequently, the general law of dopant incorporation explains not only doping but also autodoping phenomena.

References

Bloem, J., Giling, L. J., Graef, M. W. M.: J. Electrochem. Soc. Amer. **121**, 1354 (1974)
Claasen, W. A. P., Bloem, J.: ibid. **127**, 194, 1836 (1980), **128**, 1353 (1981)
Hintzen, H. T. J. M., Bloem, J., Giling, L. J.: ibid. **131**, 1900 (1984)
Kokovin, G. A., Testova, N. A., Titov, A. A., Morgenstern, Th., Kühne, H.: Cryst. Res. Technol. **20**, 1583 (1985)
Kühne, H. (I—III): ibid. **22**, Heft 4, 5, 6 (1987)
Kühne, H. (1): ibid. **20**, 479, 741 (1985), (2): Kristall u. Technik **14**, 333 (1979), (3): ibid. **14**, 413 (1979), (4): Cryst. Res. Technol. **17**, 181 (1982), (5): ibid. **17**, 1217, 1370 (1982), (6): ibid **21**, 1495 (1986)
Kühne, H., Sperling, R.: ibid. **15**, 413 (1980)
Morgenstern, Th., Kühne, H., Kokovin, G. A., Testova, N. A., Titov, A. A.: ibid. **22**, 75 (1987)
Rai-Choudhury, P., Salkovitz, E. J.: J. Cryst. Growth **7**, 353, 361 (1970)
Reif, R., Dutton, R. W.: J. Electrochem. Soc. Amer. **128**, 909 (1981)
Reif, R., Kamins, T. J., Saraswat, K.: ibid. **125**, 1860 (1978)
Wong, M., Reif, R.: IEE Transaction Electr. Dev. **ED-32**, 83 (1985)
Wong, M., Reif, R., Srinivasan, G. R.: ibid. **ED-32**, 89 (1985)

(Received December 5, 1986)

Author's address:

Dr. H. Kühne
Institut für Halbleiterphysik der Akademie der Wissenschaften der DDR
W.-Korsing-Str. 2, PSF 409
Frankfurt/Oder
DDR-1200

Cryst. Res. Technol.	22	1987	7	893–901

R. FLAGMEYER, B. SCHUMANN

Sektion Physik, WB Angewandte Kernphysik, and Sektion Chemie,
WB Kristallographie der Karl-Marx-Universität Leipzig

Characterization of Epitaxial Group II Fluoride Films on (100)-Oriented GaAs Substrates by Ion Backscattering and Channeling

Lattice matched (Ca, Sr)F_2 and related fluoride films grown on (100)GaAs substrates were investigated by Rutherford backscattering/Channeling. This technique allows the characterization of the 30–250 nm thick heteroepitaxial layers with respect to chemical composition, film thickness, interface roughness and crystalline perfection. The first flash evaporated GaAs/fluoride/GaAs structure are briefly reported.

Gitterangepaßte (Ca, Sr)F_2- und verwandte Fluorid-Schichten auf (100)GaAs-Substraten wurden mittels Rutherford-Rückstreuung/Kanalisierung untersucht. Diese Methode gestattet eine Charakterisierung der 30 bis 250 nm dicken Heteroepitaxieschichten hinsichtlich chemischer Zusammensetzung, Schichtdicke, Rauhigkeit der Interface und Kristallperfektion. Über erste mit dem Flash-Verfahren hergestellte GaAs/Fluorid/GaAs-Strukturen wird kurz berichtet.

1. Introduction

In recent years group II fluorides have received considerable attention as epitaxial insulators on semiconductors because of some potential applications and their properties, such as cubic structure, the good lattice match and the ease of evaporating stoichiometric films. Rutherford backscattering/channeling spectroscopy has been used by several authors to study perfection and epitaxial relations of heterostructures, such as GaAs or Ge/(Ca, Sr)F_2 (TU et al.; ISHIWARA et al. 1984; TSUTSUI et al.), Ge/(Ca, Sr)F_2/Si (ISHIWARA et al. 1983), Ge or InP/BaF_2 (PHILLIPS et al.), and Si/CaF_2/GaAs (ASANO et al. 1986). Films of CaF_2 on silicon substrates have been studied extensively by means of the RBS technique (e.g. ASANO et al. 1983; SCHOWALTER et al.; SMITH III et al.; HASHIMOTO et al.).

Work published in the field of mixed fluorides on III–V semiconductors has been confined to UHV evaporation and MBE growth. In this paper we present RBS/channeling investigations of (Ca, Sr)F_2 and other fluoride films and related multiple structures on (100)-oriented GaAs substrates which were deposited by the simple flash evaporation technique. TEM, RHEED, AES and RBS/channeling studies of lattice-matched $Ca_{0.43}Sr_{0.57}F_2$ films were reported briefly elsewhere (FLAGMEYER et al.; SCHUMANN et al. 1987).

2. Experimental techniques

Films with thicknesses up to 250 nm were deposited on chemo-mechanically polished GaAs substrates by the flash evaporation technique in a high vacuum apparatus (pressure of about 2 10^{-3} Pa). Powdered bulk material was evaporated in a tungsten boat at about 1825 K at typical deposition rates of 0.5 nm s^{-1}.

In the case of (Ca, Sr)F_2 films the starting material was 43 mole% CaF_2 and 57 mole% SrF_2. For this mixing ratio the lattice constant matches GaAs at R.T. The substrate temperature was varied in the range from 675 to 875 K.

The RBS/channeling investigations were performed on the van de Graaff accelerator of the Sektion Physik of the Karl-Marx-Universität using 0.8–1.6 MeV $^4He^+$ ions. The backscattering angle varied from 98° to 165° for normal ⟨100⟩ incidence, and the energy resolution of the analyzing system was about 15 keV.

3. Results and discussion

3.1. Basic characterization of single layers by RBS/channeling

Figure 1 shows backscattering spectra for two (Ca, Sr)F_2 films on (100)GaAs substrates. While the F_2 signal "rides on top" the substrate signal, the high-energy Ca edge of the thicker layer (A) and the GaAs part of the spectrum overlap. For higher values of backscattering angle and/or thinner films (sample B) the Ca edge may be separated from that of the substrate. The high mass Sr signal is in principle separated from the GaAs signal because of the kinematic factor relation $k(\mathrm{Sr}) > k(\mathrm{Ga, As})$. As illustrated by the depth scale for high He^+ energies (1.2–1.6 MeV) and large backscattering angles (120–165°) the Sr interface edge is also separated from the surface edge of Ca up to a film thickness of about 200 nm. Thus, the Sr part of the spectra appears as a "background-free" signal to characterize the film thickness and quality. The conditions for RBS analysis become most favourable for (Ca, Ba)F_2 films where the Ba signal is distinctly resolved from the GaAs step.

Random spectra were used to determine film thickness and uniformity. For energy to depth conversion, Bragg's rule was employed to get the stopping cross-section of all mixed films (Chu et al.). As schematically shown in Figure 1b, mean compositions of mixed films can be obtained in two different ways, by comparison of (i) the areas

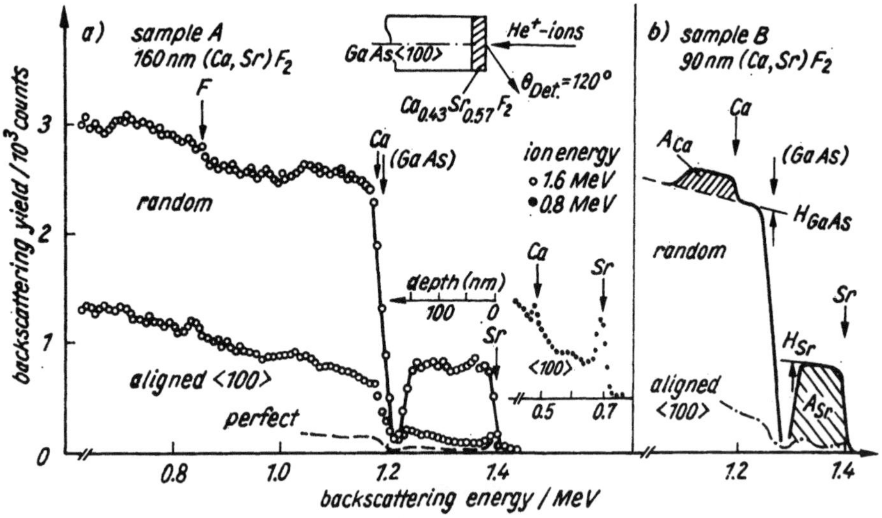

Fig. 1. Channeling and random RBS spectra of 160 nm (a) and 90 nm (b) thick (Ca, Sr)F_2 films on GaAs(100). The aligned spectrum of sample A taken with 0.8 MeV incident ion energy distinctly shows Ca and Sr surface peaks. Arrows indicate the maximum energies of He^+ ions backscattered from film components and substrate atoms

A_{Ca} and A_{Sr} in the backscattering spectrum; (ii) the step heights H_{GaAs} and H_{Sr} at the interface, without taking into account the backscattering signal from the lighter element Ca (PETROV et al.) Rutherford backscattering cross-sections of He^+ ions are given precisely for all elements and therefore target compositions can be evaluated with an accuracy of about 1%. Note that the Sr signal of sample A in Figure 1 a does not appear uniform. This means the x value of this $Ca_xSr_{1-x}F_2$ film varies slightly with depth. Except some cases in which a higher Sr content was found, the in-depth profiles were uniform and show a composition $Ca_{0.43}Sr_{0.57}F_2$ like the starting material. From aligned spectra in Figure 1 we can see that (Ca, Sr)F_2 films grow epitaxially on the GaAs substrate and the crystalline quality is good. The channeling minimum yield χ_{min} which is defined as the ratio of the aligned yield to the random yield near the surface is about 13% (sample A) and 7.5% (sample B).

The films investigated here are very thin (30–250 nm), and all information about their properties must be extracted from a small energy interval of the RBS spectra. Therefore, standard RBS conditions ($E \geqq 1.6$ MeV, $\theta \gtrapprox 160°$) are unfavourable because of the poor depth resolution. In order to improve it incident energy and the detection angle were varied. In comparison with the standard arrangement the backscattering depth factor B (cf. CHU et al.) improves from 1.5 nm/keV to ≈ 0.9 nm/keV for $\theta = 120°$ and $E = 0.8$–1.6 MeV (Fig. 1). If θ is further decreased, e.g. to 98°, the outgoing path of backscattered ions in the film increases and B becomes 0.30 nm/keV (Fig. 2). Because of the broad maximum of the stopping cross-section for (Ca, Sr)F_2 in the He^+ energy range 0.6–1.3 MeV the incident ion energy was generally held at 1.2 MeV. For the given detector resolution the surface depth resolution becomes better than 5 nm.

In this way, the surface minimum yield χ_{min} was exactly determined behind the surface peak. Minimum values of 4.5 (sample B) and 10% (sample A) for incident

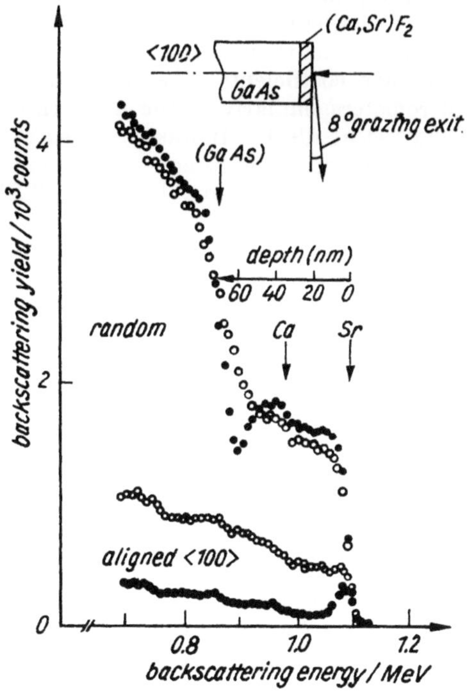

Fig. 2. High resolution backscattering spectra of thin epitaxial (Ca, Sr)F_2 films, sample C (●) and sample D (○). The depth resolution amounts to 5 nm

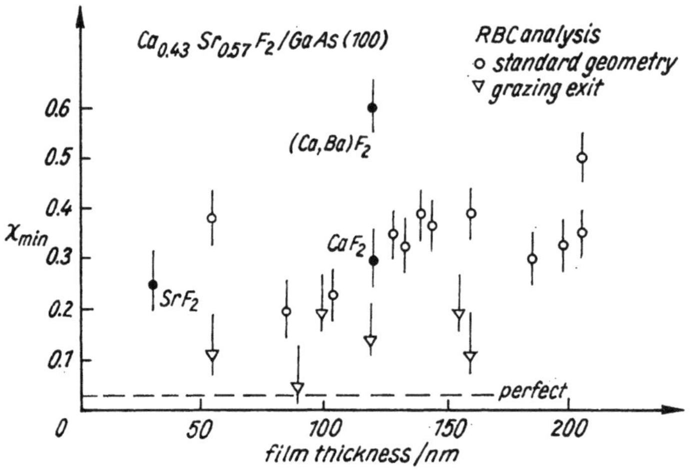

Fig. 3. Film thickness dependence of χ_{min} for all (Ca, Sr)F_2 samples grown in a substrate temperature interval from 725 to 850 K. Full circles show the best values for lattice mismatched CaF_2, SrF_2, and (Ca, Ba)F_2 films

energies of 1.2 MeV were found. Minimum yields of 3% for (Ca, Sr)F_2 on GaAs(100) (TU et al.) and CaF_2 on Si(111) (SCHOWALTER et al.) are the best results reported in the literature up to now.

In Figure 3 χ_{min} values of all samples within a substrate temperature range from 725 to 850 K for which epitaxial growth was found (SCHUMANN et al. 1986b) are summarized.

The dependence of crystalline perfection on the film thickness t exhibits a tendency toward smaller t values. This agrees well with TEM and SEM results showing cracking of films thicker than ≈ 100 nm due to strain caused by large differences in the thermal expansion coefficients. The cracks are arranged in rectangles with lengths of some micrometers similarly to the network of cracking found by SCHOWALTER et al. and SISKOS et al. 1984b. It was expected that backscattering from the "border lines" of crack patterns leads to enhanced aligned yield. On that account, χ_{min} values better than 10% are not attainable. Really, TEM images of the best sample B and of sample E with $\chi_{min} = 13\%$ not exhibit any cracking or only few cracks, respectively (GASCH).

It is noteworthy that also CaF_2 and SrF_2 grow epitaxially on GaAs in spite of large lattice misfits (CaF_2/GaAs: −3.3%, SrF_2/GaAs: 2.6%). Obviously the thermal stress is the principal cause for poor crystalline quality, and the values of lattice misfits probably play no decisive role.

The typical value of $\chi_{min} \approx 60\%$ for (Ca, Ba)F_2 results from the finegrained structure of these films. As shown by SCHUMANN et al. 1986 they consist of two epitaxially grown phases always with compositions of approximately $Ca_{0.99}Ba_{0.01}F_2$ and $Ca_{0.08}Ba_{0.92}F_2$ in accordance with the occurrence of a miscibility gap observed in the CaF_2–BaF_2 system.

3.2. Interface and surface investigations by RBS and the $^{19}F(p, p'\gamma)$ reaction

Channeling spectra are very sensitive to surface films on the epitaxial layer, as is seen in Figure 1 a (sample A) and for the grazing exit geometry in Figure 2 (sample C). We found area densities of $2.5-3.5 \cdot 10^{15}$ at. cm^{-2} Sr (corresponding to 3.5−5 mono-

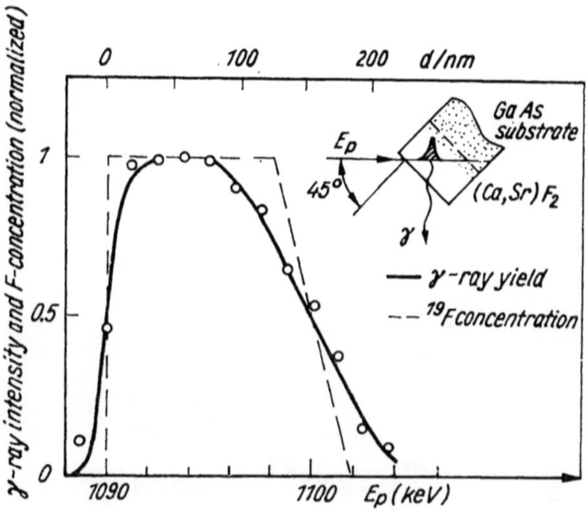

Fig. 4. Depth profiling of fluorine in a (Ca, Sr)F_2/GaAs sample using the $^{19}F(p, p'\gamma)^{19}F$ resonance at $E_p = 1090$ keV

layers) which exceed the theoretical value of a perfect single crystal. The latter arises from the interaction of an aligned ion beam with atoms in the surface monolayers even of an ideal crystal structure. As shown by FELDMAN the surface peak intensity is a universal function of one parameter $\varrho_\perp/R$. With a two-dimensional root mean square thermal vibrational amplitude $\varrho_\perp = 8.0$ pm and a shadow cone radius $R = 12.8$ pm for 1.2 MeV He^+ ions on $\langle 100 \rangle Ca_{0.43}Sr_{0.57}F_2$ rows an effective number of 1.6 atoms/row can be found, contributing to the "theoretical" surface peak. Hence there exist about 2–3.5 monolayers of disordered atoms (corresponding to 1.1 to 1.9 nm thickness) which can be understood to result from oxidation and/or drastic surface reconstruction because of the dipole moment and the large surface free energy (SCHOWALTER et al.).

The strong rise of backscattering yield in RBS spectra at the substrate edge and the Sr interface edge, as seen in Figures 1 and 2 for samples A, B, and C reflects the sharp transition from the GaAs substrate to fluoride films within a depth interval $\lessapprox 20$ nm, in agreement with AES results (SCHUMANN et al. 1987). In some cases, however, a wider transition region was found. As shown by random and aligned spectra of sample D in Figure 2 the higher interface roughness is connected with poorer epitaxial quality.

To evaluate widths of transition regions independently of the Ca or the Sr component the $^{19}F(p, p'\gamma)$ reaction was used. Because of its effective γ reaction yield and good depth resolution this resonant nuclear reaction provides a unique tool for depth profiling of fluorine (WOLF). The dashed line in Figure 4 fits the experimental points taking into account the energy widths of the beam and the resonance as well as energy straggling and shows a transition layer thickness of about 46 ± 4 nm. For a second sample, a typical value of 21 ± 3 nm was found by the same method. The more or less wide transition regions result probably from the initial roughness of the GaAs substrate (SISKOS et al. 1984b).

3.3. Depth and energy dependence of the channeling yield

Figure 5 shows depth-dependent minimum yields $\chi(d)$ for some (Ca, Sr)F_2 films with thicknesses in the range of 55 nm (sample C) to 160 nm (sample A) grown at substrate temperatures of 725 and 800 K. The high starting values $\chi(0)$ results from cracking,

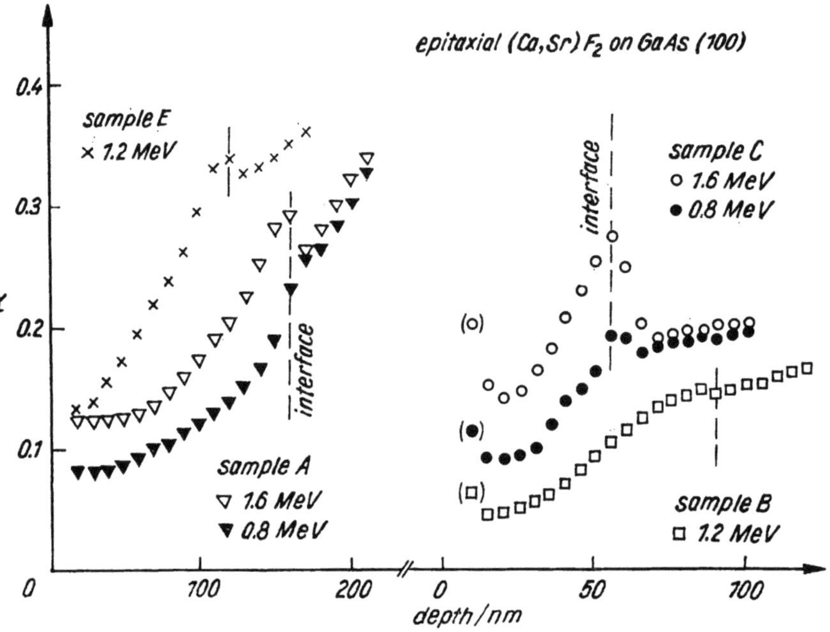

Fig. 5. Dependence of normalized channeling yield χ (aligned ⟨100⟩) on backscattering depth d. The depth resolution is 7 nm (samples B, C, E) and 14 nm (sample A). Surface peaks from Ca and Sr are omitted

as explained above. Then, defects in the whole layer lead to a more or less rapid increase of the channeling yield up to the interface. To get more information about the nature of defects we extended the evaluation of $\chi(d)$ into the substrate and investigated its dependence on the incident ion energy ($E = 0.8 - 1.6$ MeV, $\theta = 120°$). Assuming a minimum yield $\chi_p \approx 3\%$ for perfect crystals (TU et al.; KOBAYAKAMA et al.), the difference $\chi_{min} - \chi_p$ for samples A and C behind the surface peak increases linearly with E. Furthermore, from the decrease of $\chi(d)$ behind the interface as seen for samples A, C, and E in Figure 5 we conclude that direct scattering from defects contributes to the measured yield.

Recently, WIELUNSKI et al. have shown that direct scattering from dislocations should be linearly proportional to the energy of channeled ions. HASHIMOTO et al. measured the interface peak intensity of thin MBE grown CaF_2 films. The energy dependence was observed to be E^1 and was explained by considering direct scattering from curved atomic rows in the misfit dislocation network region. Taking into consideration the dechanneling fraction of ions, a variation as the square root of incident ion energy is expected when edge dislocations or loops (diameter $R > 20$ nm) are present in the films (GÄRTNER et al., KUDO). Thus we assume that relatively high dislocation densities in the films under consideration lead to strong dechanneling as well as direct backscattering of ions.

For lack of high-quality CaF_2 or mixed fluoride crystals $\chi_p(d)$ was determined by a theoretical χ_{min} value of 3% at the surface and a small increase with depth based on the 2 MeV He^+ measurements on perfect (111) CaF_2 crystals (SKOG) in view of the strong dechanneling by monatomic F rows along the ⟨100⟩ direction. Finally, from the increase of $\chi(d) - \chi_p(d)$ with depth we conclude a high dislocation density between 10^{10} cm^{-2} (samples A, B) and 10^{11} cm^{-2} (C, E) in comparison with results from Si

(GÄRTNER et al.) and CaF_2 (SCHOWALTER et al., SMITH III et al.) investigations. No single defects could be resolved in TEM images even from our best samples (SCHUMANN et al. 1987).

It should be noted that polycrystalline structures also exhibit an increase of χ_{min} with increasing energy (SIGURD et al.). Provided that the films in question are normal-textured polycrystallites the spread of crystallite orientations $\Delta\psi$ for samples A and D would be 0.1—0.2°, whereas χ_{min} values of 20—40% would be related to quantities of about 0.3—0.6°. However, no polycrystalline structure was detected by TEM or RHEED, and the strong increase of $\chi(d)$ appears untypical of thin polycrystalline films.

3.4. GaAs-fluoride film systems

The growth of GaAs/insulator/GaAs structures was carried out in the same flash-evaporating system without breaking the vacuum. Figure 6a shows backscattering spectra of a GaAs/CaF_2/GaAs (100) structure grown at 825 K. The aligned spectrum indicates that the GaAs film grows epitaxially on the CaF_2/GaAs system though the quality of the underlaying film is poor. For comparison, sample 2 without the GaAs top layer exhibits a channeling minimum yield χ_{min} at the CaF_2 surface of 30% (see inset). The χ_{min} of the GaAs film is very high at the interface and decreases to about 45% at the surface, indicating that the crystalline quality of the heteroepitaxial GaAs is improved with increasing thickness.

This agrees well with the results of ASANO et al. (1986). MBE grown 410 nm thick GaAs films on (100) oriented CaF_2/Si structures show χ_{min} values in the range of 15 to 45% and 30—50% at substrate temperatures of 725 and 825 K, respectively.

One interesting feature of the dependence of the interface width on the growth sequence is that the transition in the CaF_2/GaAs (bulk) system is sharper than that in the

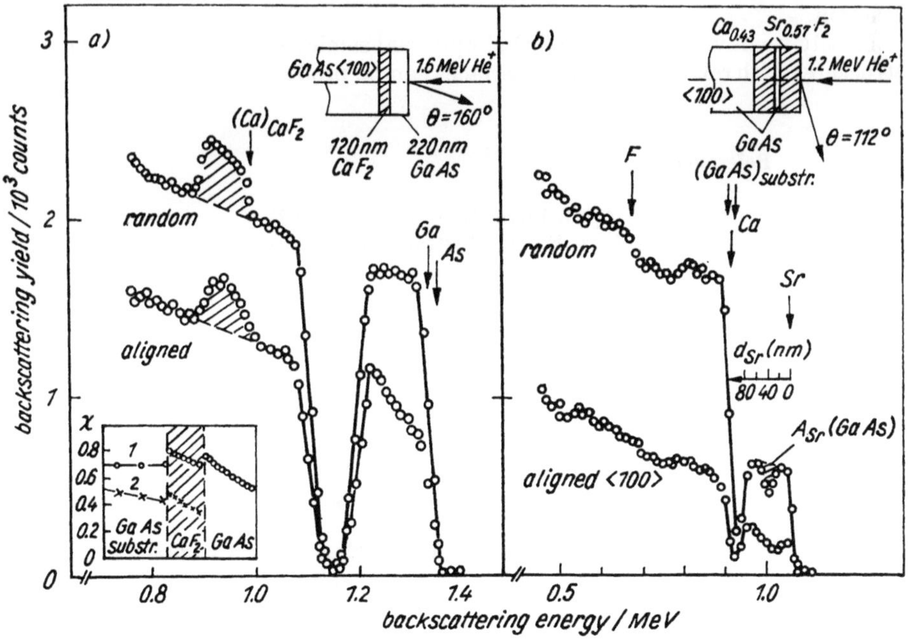

Fig. 6. RBS/Channeling spectra of a GaAs/CaF_2/GaAs(100) structure (a) and a (Ca, Sr)F_2/GaAs/(Ca, Sr)F_2/GaAs(100) structure (b)

GaAs (layer)/CaF_2 system. Taking into account the depth resolution of the analytical method for GaAs (including the energy straggling of He^+ ions) one can deduce from the random spectrum widths of transition regions of 15 nm and 35 nm for the first and second interfaces, respectively. Epitaxial growth of GaAs on (Ca, Sr)F_2/GaAs (100) structures which was demonstrated for the first time by SISKOS et al. (1984a) was studied for different thicknesses of GaAs films and lattice matched $Ca_{0.43}Sr_{0.57}F_2$ films. Figure 6b shows random and aligned backscattering spectra of two epitaxial fluoride layers on GaAs substrate with an intermediate GaAs film grownat 800 K. The crystalline quality of this (Ca, Sr)F_2/GaAs/(Ca, Sr)F_2/GaAs (100) structure is good (χ_{min} = = 23%) for the fluoride top layer. The dip in the Sr signal at 1.0 MeV arises from the very thin semiconductor film. The dip area A_{Sr} (GaAs) is related to a GaAs layer of 7 nm thickness, whereas from the energy width it follows that theGaAs film is extended over about 15 nm. As expected, the interfaces are not atomically smooth, but the semiconductor film appears distinctly separated from the isolator material.

4. Summary

It was shown that the Rutherford backscattering and channeling technique allows the characterization of thin fluoride films on GaAs substrate in respect of composition, thickness and crystalline perfection. The best values of channeling minimum yield χ_{min} of 4.5% and 10% are found for film thicknesses below 100 nm. The stress introduced by the large thermal mismatch between GaAs and fluoride causes cracking as well as high dislocation densities, which was proved by ion dechanneling and direct backscattering. Amorphous surface layers were evaluated with a resolution of 0.5 monolayers, and the extents of interface regions in the range of about 20 nm were determined advantageously by means of the RBS technique and the $^{19}F(p, p'\gamma)$ reaction.

Finally we have demonstrated that epitaxial semiconductor/fluoride/semiconductor structures with relatively sharp interfaces can be produced by flash evaporation.

References

ASANO, T., ISHIWARA, H.: Appl. Phys. Lett. **42**, 517 (1983)

ASANO, T., ISHIWARA, H., LEE, H. Ch., TSUTSUI, K., FURUKAWA, S.: Jap. J. Appl. Physics **25**, L139 (1986)

CHU, W. K., MAYER, J. W., NICOLET, M. A.: Backscattering Spectrometry, New York 1978

FELDMAN, L. C.: Appl. Surface Sci. **13**, 211 (1982)

FLAGMEYER, R., SCHUMANN, B.: phys. stat. sol. (a) **93**, K117 (1986)

GASCH, W.: Diploma work, Karl-Marx-Universität, Leipzig 1985

GÄRTNER, K., HEHL, K., SCHLOTZHAUER, G.: Nucl. Instrum. Meth. Phys. Res. **B4**, 63 (1984)

HASHIMOTO, S., PENG, J. L., GIBSON, W. M., SCHOWALTER, L. J., FATHAUER, R. W.: Appl. Phys. Lett. **47**, 1071 (1985)

HASHIMOTO, S., WIELUNSKI, L. S., PENG, J. L., GIBSON, W. M., SCHOWALTER, L. J.: Nucl. Instrum. Meth. Phys. Res. **B13**, 65 (1986)

ISHIWARA, H., ASANO, T.: Jap. J. Appl. Physics **22**, Suppl. 22-1, 201 (1983)

ISHIWARA, H., TSUTSU, K.I, ASANO, T., FURUKAWA, S.: Jap. J. Appl. Physics **23**, L803 (1984)

KOBAYAKAMA, M., NIWA, H., TANEMURA, S.: J. Phys. Soc. Jap. **36**, 207 (1974)

KUDO, H.: Phys. Rev. **B18**, 5995 (1978)

PETROV, I., BRAUN, M., FRIED, T., SÄTHERBLOM, H. E.: J. Appl. Physics **54**, 1358 (1983)

PHILLIPS, J. M., FELDMAN, L. C., GIBSON, J. M., MCDONALD, M. L.: J. Vac. Sci. Technol. B1, 246 (1983)
SCHOWALTER, L. J., FATHAUER, R. W., GOEHNER, R. P., TURNER, L. G., DEBLOIS, R. W., HASHIMOTO, S., PENG, J. L. GIBSON, W. M., KRUSIUS, J. P.: J. Appl. Physics **58**, 302 (1985)
SCHUMANN, B., KÜHN, G., WAGNER, G.: Cryst. Res. Technol. **21**, 1139 (1986)
SCHUMANN, B., KÜHN, G., GASCH, W., WAGNER, G., FLAGMEYER, R., MÜLLER, O., HAUFFE, W.: J. Cryst. Growth **82**, 405 (1987)
SIGURD, D., BOWER, R. W., VAN DER WEG, W. F., MAYER, J. W.: Thin Solid Films **19**, 319 (1973)
SISKOS, S., FONTAINE, C., MUNOZ-YAGUE, A.: Appl. Phys. Lett. **44**, 1146 (1984) a
SISKOS, S., FONTAINE, C., MUNOZ-YAGUE, A.: J. Appl. Physics **56**, 1642 (1984) b
SKOG, G., LINDEN, M., HELLBORG, R.: Rad. Eff. **37**, 59 (1978)
SMITH III, T. P., PHILLIPS, J. M., PEOPLE, R., GIBSON, J. M., STILES, P. J.: Mat. Res. Soc. Symp. Proc. **37**, 163 (1985)
TSUTSUI, K., ISHIWARA, H., FURUKAWA, S.: Appl. Phys. Lett. **48**, 587 (1986)
TU, C. W., WANG, S. J., PHILLIPS, J. M., GIBSON, J. M., STALL, R. A., WUNDER, R. J.: J. Vac. Sci. Technol., in press
WIELUNSKI, L. S., HASHIMOTO, S., GIBSON, W. M.: Nucl. Instrum. Meth. Phys. Res. **B13**, 61 (1986)
WOLF, U.: Thesis, Karl-Marx-Universität, Leipzig 1985

(Received December 12, 1986)

Authors' addresses:

Dr. R. FLAGMEYER
Sektion Physik, WB Angewandte Kernphysik,
Karl-Marx-Universität
Linnéstraße 5
Leipzig
DDR-7010

Dr. B. SCHUMANN
Sektion Chemie, WB Kristallographie,
Karl-Marx-Universität
Talstraße 35
Leipzig
DDR-7010

Book Review

A. B. P. Lever (ed.)

Excited States and Reactive Intermediates

ACS Symposium Series, Vol. 307

American Chemical Society, Washington, DC, 1986, Price $ 68.95, 276 pp., ISSN 0097-6156, 307, ISBN 0-8412-0971-5

Der vorliegende Band, der seit 1974 erscheinenden Monographiereihe „ACS Symposium Series", enthält die wesentlichsten Beiträge des „1985 Biennial Inorganic Chemical Symposium", Toronto/Canada, das gemeinsam von der American Chemical Society und dem Chemical Institute of Canada veranstaltet wurde und sich mit neueren Resultaten und Ergebnissen zur Photochemie und -physik angeregter Zustände, ihrer experimentellen und theoretischen Zugänglichkeit beschäftigte. Seit der Erkenntnis, daß das Molekül im angeregten Zustand eine im Vergleich zum Grundzustand eigenständige, neue Spezies darstellt (A. W. Adamson, 1975), die in den meisten Fällen verschiedene Struktur, Stereochemie, verschiedene Elektronen- und Schwingungsspektren, kurzum eine vom Grundzustand verschiedene Chemie aufweist, wird dieses Feld sehr intensiv untersucht. Die transiente Natur der angeregten Zustände war die Ursache, daß besonders mit der Einführung der Lasertechnik und elektronischer Geräte mit extrem schneller Registrierung ein enormer Anstieg der Forschungsaktivitäten zu verzeichnen ist.

Die ausgewählten 16 Beiträge stellen besonders originelle, wissenschaftlich anspruchsvolle Arbeiten zur Natur und Reaktivität dieser kurzlebigen Spezies dar. Einkristall-Elektronenspektren bei tiefen Temperaturen (H. U. Güdel), Ab-initio-Analysen von CT-Übergängen (L. G. Vanquickenborne) und Lebensdauern des niedrigsten angeregten Dublett-Zustandes (J. F. Endicott) an Cr(III)-Komplexen werden in drei Kapiteln dargestellt. Mit Aussagen zur Geometrie angeregter Zustände von Koordinationsverbindungen beschäftigen sich Arbeiten von H.-H. Schmidtke und J. I. Zink. Neben der Elektronenspektroskopie ist vor allem die Resonanz-Raman-Spektroskopie in der Lage, detaillierte Informationen über die elektronische und Schwingungsnatur angeregter Zustände zu liefern. So stellen R. J. H. Clark RR-spektroskopische Studien an eindimensionalen Systemen (linear verkettete Pt-Komplese) und M. J. Weaver Raman-Untersuchungen adsorbierter Reaktanten an Elektroden vor. Weitere Beiträge beschäftigen sich mit dem Problem der statischen spektralen Sensibilisierung photokatalytischer Systeme (H. Hennig), mit dem biologisch relevanten Problem des Elektronentransfers über große Entfernungen in Proteinen (G. McLendon), der elektrochemischen Erzeugung von Metallkomplexen in angeregten Zuständen (A. Vogler), der Photochemie von Metall-Metall-gebundenen Zweikernkomplexen (D. J. Stufkens u. J. L. Marshall) sowie der photochemischen Erzeugung reaktiver organometallischer Spezies (W. C. Trogler).

Es kann festgestellt werden, daß die vorliegende Monographie eine gelungene Auswahl von Fortschrittsberichten, verfaßt von den führenden Vertretern des Forschungsgebietes, bietet, die einen vorzüglichen Überblick über den gegenwärtigen Stand eines sich rasant entwickelnden Teilgebietes der Chemie gestattet.

R. Benedix

Cryst. Res. Technol.	22	1987	7	903–909

B. Wolf, A. Zehe

Sektion Physik, Technische Universität Dresden

Are In-growth Surface Reactions during MBE Controlled by a "Quasi-Gas Transition Layer"?

Recently a new growth model for Molecular Beam Epitaxy (MBE) was presented in literature (Herman, Andersson), that assumes a gas-like transition layer between the ordered crystal and the vacuum. Known experimental results fit the picture of a disordered layer on top of a growing crystal which reacts much quicker with residual gas constituents than the static surface does. Following this idea, consequences of the "Near Surface Quasi-Gas" model (NSQG model) with respect to increased chemical reactivity during MBE growth are discussed in the present paper.

1. Introduction

Molecular Beam Epitaxy (MBE) is a crystal growth technique which proceeds far from equilibrium conditions and is governed more by the kinetics of incoming particle fluxes than by the thermodynamic laws. Thus metastable compounds have been produced quite easily. Nevertheless MBE growth is not fully understood and still subject to active exploration. Recently a growth model has been proposed that assumes a "Near surface quasi-gas" (NSQG) transition layer on top of the growing crystal Herman, Andersson; Herman et al. 1986).

It is taken that this layer, which is up to 4 monolayers thick, can be compared to a thin liquid film with respect to its density, and to a gas with respect to the degree of order and the mobility of the molecular or atomic species. Though beam equivalent pressures in the range of 10^{-4} Pa, typical for MBE, seem to support the picture of a sharp transition between crystal and vacuum, some experimental results as given below suggest the existence of a NSQE: (i) precursor states in the adsorption of As_2 and As_4 species during GaAs–MBE, as found by modulated molecular beam experiments (Foxon, Joyce 1975, 1977). (ii) Intensity variations in RHEED pattern, at first observed in 1983 (van Hove et al.) are corresponding to a 2-dimensional growth. The damping indicates the beginning of a 3-dimensional growth that is according to Herman et al. due to the formation of a NSQG. (iii) Changes in reevaporation rates. A drastical change in the reevaporation rates of Cd and Te in the case of CdTe atomic layer epitaxy (ALE) was observed near the substrate-deposit interface (Herman et al. 1986). These evaporation rates, by 2 orders of magnitude smaller than those of bulk-like amorphous Cd and Te solids were interpreted in terms of the formation of a NSQG. (iv) Successful optimization of ALE process under the assumption of a NSQG formation (Herman et al. 1985). (v) The Potential Enhanced Doping (PED) concept of Si–MBE may be understood in terms of the existence of a NSQG (Kubiak, Parker).

In our opinion the existence of a NSQG is decisive for the increase in chemical reactivity of a growing MBE surface. Studies devoted to O_2- and H_2 adsorption on perfect in-situ cleavage planes of semiconductors (Kreutz et al.; Mönch; Proix et al.;

ANTONANGELI et al.) revealed a very weak affinity towards molecular O_2 and H_2. On the other hand post-growth examinations of impurity incorporation (CALAWA) and in-growth observations of growth modification by hydrogen (BACHRACH, BRINGANS) indicate a considerably increased yield of chemical reactivity which is consistent with the formation of a NSQG. This paper presents some results of SIMS- and AES measurements, devoted to oxygen and hydrogen adsorption on ion sputtered InSb surfaces. Thereby the assumption is made that ion damaged surfaces can, to some extend, simulate the chemical reactivity of a growing MBE surface adjacent to a NSQG. Furthermore, consequences of the NSQG with respect to impurity incorporation from the residual gas atmosphere in the UHV chamber are discussed.

2. Experimental

The InSb samples under study were obtained by ex-situ cleaving an InSb single crystal ingot grown by the horizontal Bridgman technique. The samples were inserted into the UHV chamber without further mechanical or chemical treatment. In-situ ion sputtering was carried out which resulted in a clean surface rich in sputtering induced defects. Sputter roughening and a high density of imperfections caused an increased surface reactivity that enabled us to study the adsorption of both H_2 and O_2 at medium large exposures already in the range of some 10^2 to 10^3 Langmuirs.

The analyzing equipment involved both AES- and SIMS facilities (CMA with coaxial electron gun and Ar^+-ion gun together with a RIBER-QMS, respectively). Research-grade pure gases in glass tubes were used for the adsorption experiments. To study the temperature influence on adsorption, the sample holder was equipped with a resistance tungsten heater.

3. Results and discussion

3.1. Adsorption of oxygen

The typical development of oxygen coverage with increasing oxygen exposition dose for an ion bombarded InSb (110) surface is depicted in Figure 1. Since immense differences in reactivity between molecular and atomic or excited oxygen are reported in literature (PIANETTA et al.), the adsorption was carried out in a first run with the ion pump working, and also without ion pumping. No significant differences were found, which is interpreted in terms of a high cracking efficiency for molecular species of the imperfect surface itself. If compared to adsorption experiments on perfect in-situ

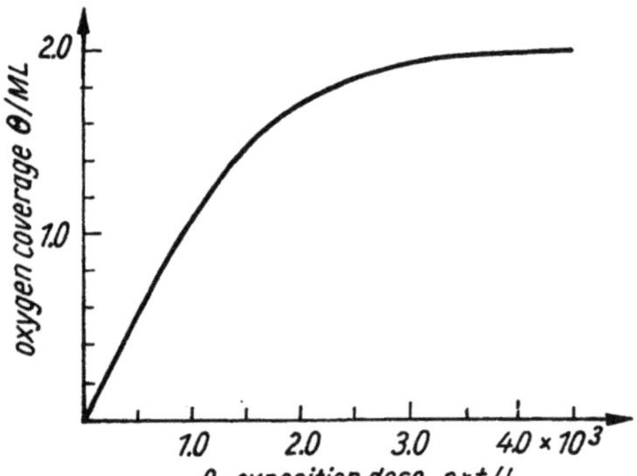

Fig. 1. Oxide formation as a function of molecular oxygen exposure

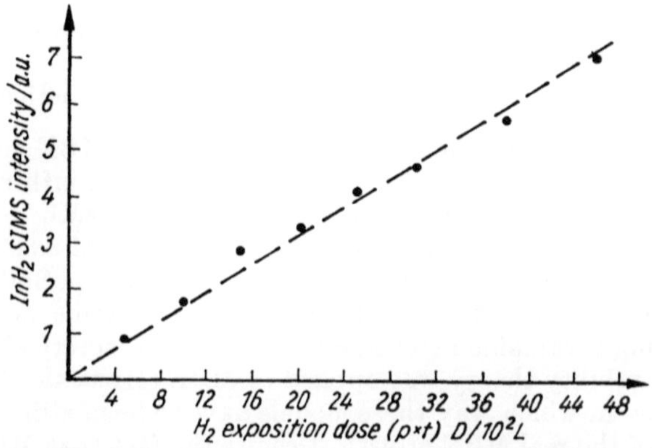

Fig. 2. SIMS-signal height of InH_2^+-clusters, normalized to the In^+-signal, as a functionof molecular hydrogen exposition (read InH_2^+)

cleavage planes as mentioned before the following differences are obvious: (i) Surface reactivity is increased by a factor of 10^3 to 10^4 (about 10^3 L O_2 form 1 monolayer of oxide, in contrast to 10^6-10^7 reported in literature). (ii) The oxide layer thickness corresponding to the saturation value (about 2 monolayers) is by a factor of 3 larger than in the case of a perfect surface (0.7 monolayers).

3.2 Adsorption of hydrogen

A detailed description of these experiments is given elsewhere (Wolf, Zehe). Only the main results are summarized here: SIMS spectra revealed hydrogen containing clusters, namely InH^+, InH_2^+, $InKH^+$, $InOH^+$, In_2OH^+. Figure 2 illustrates the increase of the InH_2^+-signal with growing hydrogen exposition dose. SIMS spectra suggest that hydrogen exists in both physisorbed and chemisorbed stages.

Compared to in-situ cleavage planes of $A^{III}B^V$ semiconductors a considerably increased reactivity power towards hydrogen was found.

3.3. Discussion

The interaction of molecular hydrogen with the surface is supposed to proceed in two steps: physisorption, followed by dissociative chemisorption. A direct transition into a chemisorbed stage similar to findings for GaAs and Si (Bartels et al.; Bacharach et al.; Schulze, Henzler) seems possible only in case of atomic hydrogen. The same should apply for oxygen.

The sputter-induced surface roughness results in an extended effective surface area and a high number of crystal defects. Owing to the roughness structure, strong local fields will exercise electrical forces and that enhanced of the van-der-Waals kind on impinging molecules. This in combination with the increased number of adsorption sites will give rise to a considerably increased sticking probability in a physisorbed state. While physisorption may well be a collective result of surface imperfections as a whole, the subsequent dissociation into reactive single atoms is due to the action of an individual, localized imperfection on the mobile physisorbed molecule. The resulting atoms will either be bound to imperfections, that have caused the dissociation process, or may migrate over the surface until they can saturate free valences (dangling bonds). The latter situation can be probed by means of modulated molecular beam mass spectroscopy: By offering D_2 and H_2 molecules to the surface the desorption products should contain HD species, if H and D atoms on the surface recombine after migration.

4. The model of the NSQG and its effect on MBE growth

4.1. Consequences for growth kinetics

By accepting the concept of an NSQG the MBE growth proceeds as follows: i) Formation of the NSQG by a continuous particle influx, — a process far from equilibrium. ii) Solidification and growth out of the NSQG, — a process that involves features of both gas-phase epitaxy and liquid-phase epitaxy and which is not any more far from equilibrium conditions. The latter process permits the application of well developed models and theories to MBE, that describe near-equilibrium processes, representing a valuable extension of applicable theoretical tools. In a first approach the linear stability theory was applied to MBE (HERMAN, ANDERSSON). The question, however, is, in which way these models have to be modified in order to get a proper description of the real MBE-growth. Despite the fact that MBE is characterized by a high supersaturation (the ratio between impinging and equilibrium desorption fluxes), the growth-model of Burton, Cabrera and Frank (BCF theory) — established for low supersaturation — is also applied to MBE (FUENZALIDA). After BCF the impinging species condense mainly into step edges at low fluxes. A smooth layer-by-layer growth proceeds, resulting in high crystalline perfection.

With increasing fluxes the number of adatoms, of collisions and therefore of nucleation centers raise. Nucleation becomes a competitive process; consequently island growth will proceed. This is accompanied by a decrease of crystal perfection. Three-dimensional growth is unavoidable at high fluxes.

From the existence of a NSQG the following modifications arise: At a given flux the population density of adatoms is much higher than without NSQG. Nucleation is therefore of importance from the very beginning. Solidification proceeds as a 2-dimensional island growth. Since the number of nucleation centers is large the maximum diameter of the islands before coalescence (fusion of islands) is small. The strong interaction between the substrate, the islands and the NSQG results in a softening (muggyfying) of the solidification front. This allows easily rearrangements of atoms with respect to the substrate during and after island fusion. At a given island density the number of crystal imperfections after growth is expected to be much lower than in the corresponding case without a NSQG.

Thus, the NSQG concept is able to explain the high quality of MBE-layers even in case of comparatively high growth rates (10 µm/h), as e.g. in Si–MBE.

4.2. The role of an NSQG for the incorporation of residual gas components

The strong interaction between particles within the NSQG and with the substrate as well as the softening of the solidification front suggests an increase in chemical reactivity to residual gas components. As the initial stages of oxidation have been shown to be strongly attached to imperfections (KREUTZ et al.; MÖNCH et al.), the impurity incorporation may proceed via transient imperfections, that are present during growth. Provided that the enhanced island formation under the influence of the NSQG, as proposed above is a correct concept the total length of step boundaries is considerably longer than under growth conditions without additional nucleation. This would explain an increase of impurity incorporation. Still of more importance is the fact, that molecular impurities (e.g. O_2, H_2O, H_2) that reach the step edges or island boundaries in a stage bound to small clusters instead of single molecules are incorporated readily. The bonding to such clusters can occur either on the substrate or in the NSQG. It is evident that the probability to do so depends both on the cluster density and on the residence time an impurity spends on the solidification border or in the NSQG, respectively. The cluster density is a function of the adatom density and grows mainly with the square of the latter. The residence time is also increased as compared to the BCF model. In contrast to the BCF growth mode, where short-time adsorbed impurity molecules leave the surface for the vacuum, an exchange between adjacent layers occurs in case of the presence of the NSQG. This means that the mere existence of a NSQG explains the incorporation of impurities from the growth environment to the actually measured higher concentration level.

There is a physical similitude between a growing MBE interface and a static but imperfection rich crystal surface. Contemporary MBE equipments guarantee a base pressure below 10^{-8} Pa, while the partial pressure of undesired components, as e.g. CO, CO_2, O_2 is expected to be in the 10^{-9} Pa range. By assuming a sticking coefficient of 10^{-7} as reported for oxygen on cleavage planes, an MBE layer grown at a rate of 1 monolayer/s should display an impurity level of about 10^{11} cm^{-3}. But in fact this level commonly exceeds an amount of 10^{14} cm^{-3}, which indicates a sticking probability in the 10^{-3} to 10^{-4} range (Pamplin, Fuenzalida). This value agrees quite well with the sticking coefficient that was found in case of ion-damaged surface as presented in this paper. Thus, it seems plausible that the surface roughness, induced by ion bombardment, and the "transient roughness" during growth behave similarly with respect to impurity adsorption enhancement. Furthermore ion bombardment results in deviations from stoichiometry. For instance, the group-III-elements are commonly enriched in case of III–V compounds. The excess component might form clusters with high affinity to oxygen and other gases, which are similar to those, over which the oxygen incorporation during growth proceeds. Finally the situation characterized by missing bulk-like bonds on the one hand and strong interaction between the species of the NSQG including the "softening" of the crystallization front on the other hand is to some extend comparable with the situation of broken bonds in case of ion damaging.

4.3. Molecular beam flux-dependence of impurity sticking

The sticking behaviour of impurities on solid surfaces depends on the state of crystalline perfection. From this view a difference of sticking coefficients for static cleavage planes and dynamic growing crystal faces is likely to happen. Proceeding on this assumption, the total impurity sticking coefficient on a MBE growing film is written in the form $\alpha_i = \alpha_{i,0} + \alpha_{i,g}(j)$ where $\alpha_{i,0}$ describes the sticking behaviour on the static surface, while $\alpha_{i,g}(j)$ involves all growth-dependent processes, that contribute to impurity incorporation; j is the incoming flux. Particularly the following aspects are of importance here:

a) Moving step edges and island formation

Under static conditions an impurity molecule with low sticking probability that impinges on the crystal surface will migrate during a characteristic time τ on the surface, reaching a maximum distance $R_m(\tau)$ from the initial departure point. It will be adsorbed provided there is an appropriate point defect or step edge within the circle of diameter $2R_m$ that permits chemisorption. At small growth rates the BCF growth mode is applicable to MBE growth, based on moving step edges. The probability that an impurity particle will reach a step edge may be increased on the growing surface.

Additional step edges arise as island borders during island formation and growth. The total length of island borders is linked by the same arguments to the contribution in sticking enhancement.

b) Incorporation via sticking to clusters

The sticking of an impurity to a cluster and subsequent common incorporation onto the growing crystal surface seems to be a rather efficient process under NSQG transition layer conditions. Let cluster formation start by the collision of two particles within the NSQG, for simultaneous collisions involving higher particle numbers are of considerably less importance. Of course, matrix-building clusters (containing only intentionally introduced elements) will always outnumber those clusters which contain also impurity atoms. But as a consequence of the formation mechanisms of the clusters, a relation $n_{cl} \sim n_{ad}^2$ between the cluster density and the adatom density is established. Since n_{ad} is proportional to the incoming flux j according to $n_{ad} = \tau \cdot j$ (τ: mean residence time until desorption or incorporation occurs), the number of incorporated impurities via such a process will at least depend on the square of the incoming flux or, — what is equivalent, on the square of the growth rate.

It is a task of future experimental work to detail the weight of processes responsible for the observed impurity incorporation enhancement.

5. Summary

It has been experimentally shown that the sticking coefficients for impurities on semiconductor surfaces during MBE growth are by several orders of magnitude higher than at static surfaces, and are thus comparable with those for ion-damaged surfaces. The NSQG-concept was used to physically explain these findings. The modification of BCF growth was shown to result in an increased nucleation and island growth, but under the softening influence of the NSQG both rearrangements of atoms are favoured according to the crystallography of the underlayer, and the resulting imperfection level is reduced. Step motion, island formation and adsorption onto small clusters have been shown to enhance impurity incorporation. A quantitative discussion of the contribution of these processes will require detailed experimental data.

One of us (B.W.) wants to express his gratitude to Dr. Schuricht and co-workers, Humboldt-Universität Berlin, for providing the InSb samples used in this study.

References

Antonangeli, F., Calandra, C., Colavita, E., Nannarone, S., Rinalid, C., Sorba, L.: Phys. Rev. **B29**, 8 (1984)

Bachrach, R. Z., Bringans, R. D.: J. Vac. Sci. Technol. **B1**, 142 (1983)

Bartels, F., Surkamp, L., Clemens, H. J., Mönch, W.: J. Vac. Sci. Technol. **B1**, 756 (1983)

Calawa, A. R.: Appl. Phys. Lett. **33**, 1020 (1978)

Foxon, C. T., Joyce, B. A.: Surf. Sci. **50**, 434 (1975), **64**, 293 (1977)

Fuenzalida, V.: Doctoral Thesis, München 1985

Herman, M. A. Andersson, T. G.: Appl. Phys. **A41**, 1 (1986)

Herman, M. A., Vulli, M., Pessa, M.: J. Cryst. Growth **73**, 403 (1985)

Herman, M. A., Jylhä, O., Pessa, M.: Cryst. Res. Technol. **21**, 841, 969 (1986)

Kreutz, E. W., R ickus, E., Sotnik, N.: Surf. Sci. **151**, 52 (1985)

Kubiak, R., Parker, E.: Electronics & Power, Nov./Dec. 1984, pp. 853

Madukhar, A.: Surf. Sci. **132**, 344 (1983)

Mönch, W.: Surf. Sci. **168**, 577 (1986)

Neave, J. H., Joyce, B. A., Dobson, P. J., Norton, N.: Appl. Phys. **A31**, 1 (1983), **A34**, 1779 (1984)

Pamplin, B. R. (editor): Molecular Beam Epitaxy, New York 1980
Pianetta, P., Lindau, I., Garner, C. M., Spicer, W. E.: Phys. Rev. **B18**, 2792 (1978)
Proix, F., M'Hamedi, O., Sebenne, C. A.: Proc. ECOSS-7, 141 (1985), Marseille
Schulze, G., Henzler, M.: Surf. Sci. **124**, 336 (1983)
Wolf, B., Zehe, A.: Submitted for publication to "Thin Solid Films"
van Hove, J. M., Lent, C. S., Pukite, P. R., Cohen, P. J.: J. Vac. Sci. Technol. **B1**, 741 (1983), **B3**, 563 (1985)

(Received November 29, 1986)

Authors' address:

Prof. Dr. Dr. A. Zehe, DP. B. Wolf
Sektion Physik, Technische Universität Dresden
Mommsenstraße 13
Dresden
DDR-8027

Cryst. Res. Technol.	22	1987	7	910

Rook Review

A. LAUBEREAU, M. STOCKBURGER (eds.)

Time-resolved vibrational spectroscopy

Proceedings of the 2nd international conference, EMIL-WARBURG-Symposium, Bayreuth-Bischofsgrün, Fed. Rep. of Germany, June 3–7, 1985. (Volume 4 of series Springer Proceedings in Physics)

Springer-Verlag Berlin, Heidelberg, New York, Tokyo 1985, XI + 302 pages, 187 figures, DM 88.–, ISBN 3-540-16175-9

The greater number of readers will probably feel surprized to find a book with this title reviewed in this journal. Time-resolved vibrational spectroscopy as a newer field of research was started with investigations in the gaseous state and expanded quickly into the field of liquids. From this volume we learn in an impressive way that also applications to the solid state are of high innovative power.

To be sure, „Vibrational Dynamics in Solids" is only one chapter between six further ones. But applications to high polymers and to organized biological systems are not very distant from those to crystals. The contributions grouped under „Techniques and Instrumentation" and „Theoretical Aspects" are of relevancy regardless of state.

With such conference proceedings comprizing sixty contributions of 4.5 pages as the mean the question arises as to what extent a coherent representation of the field is provided by the book. In this regard it is worthy of appreciation that most contributors took care to introduce within their limited space the reader sufficiently to their special subject. Others were confident that their great name would be enough to show the importance of their work. The great care taken by the editors also when they drew up the programme and when they instructed their speakers is really felt. The book was assembled from the camera-ready manuscripts of the contributors. By means of a generous interspersing with captions from letter-press types the producer achieves a clearness of arrangement which with more simple products of offset printing generally will be missed. This volume leaves from its contents, from its arrangement and from its appearance a very pleasant impression.

W. E. STEGER

Cryst. Res. Technol.	22	1987	7	911–922

A. HUPFER, S. SCHULZE, D. HIRSCH, L. ŻDANOWICZ

Sektion Physik der Karl-Marx-Universität, and
VEB Werk für Fernsehelektronik Berlin, DDR, and
Zakład Fizyki Ciała Stałego, Polska Akademia Nauk, Zabrze, Poland

Surface Preparation of $A_3^{II}B_2^{V}$ Semiconductors (Cd_3As_2, Zn_3As_2, Cd_3P_2, Zn_3P_2) under Ultra-High Vacuum Conditions

The UHV surface preparation of $A_3^{II}B_2^{V}$ materials (crystals and thin films) has been monitored with XPS and AES. Clean and stoichiometric surfaces of $A_3^{II}B_2^{V}$ crystals were prepared by means of low energy ion bombardment and subsequent low temperature annealing. Stoichiometric Cd_3As_2 and Zn_3P_2 thin films with very low amounts of C and O were deposited by the evaporation of bulk material in the UHV. The quality of prepared $A_3^{II}B_2^{V}$ crystal and thin film surfaces was sufficient to carry out density of states investigations (UPS, RELS) with success.

Приготовление поверхностей $A_3^{II}B_2^{V}$ образцов (кристаллов и тонких слоев) в ультравысоком вакууме (УВВ) было контроллировано с помощью РФЭС и ОЭС. Проготовление чистых и стехиометрических поверхностей $A_3^{II}B_2^{V}$ кристаллов было сделано с помощью бомбардировки ионами малой энергии и допольнительным рекристаллизационным отжигом. Тонкие стехиометрические слои Cd_3As_2 и Zn_3P_2 с особо малой концентрацией примесей (С и О) были сублимированы посредством испарения исходного материала в УВВ. Качество приготовленных поверхностей $A_2^{II}B_2^{V}$ кристаллов и тонких слоев было достаточно чтобы провести эксперименты по плотности электронных состояний (УФЭС, СХПЭЭ) с успехом.

1. Introduction

The group-II^B and V^A elements of the periodic system form various types of compounds with interesting combinations of semiconducting properties. The structural, optical, transport and other properties especially of the $A_3^{II}B_2^{V}$ family of zinc and cadmium with phosphorus and arsenic have been intensively studied in recent years (see for overview W. ŻDANOWICZ; GELTEN, L. ŻDANOWICZ; LAZAREV et al.). Zn_3P_2 has been investigated as one of the most promising semiconductors in view of low-cost and high-efficiency terrestrial solar-cell applications (e.g. PAWLIKOWSKI). Some applications of Cd_3As_2 were suggested in connection with its unusual transport properties (summarized by W. ŻDANOWICZ). All $A_3^{II}B_2^{V}$ semiconductors are used as diffusion sources for doping $A^{III}B^{V}$ materials with group-II^B elements (MATSUMOTO).

We have carried out electron spectroscopic investigations such as UV and X-ray photoemission (UPS/XPS) and low energy electron loss spectroscopy in the reflection mode (RELS) which provide information on the energy band structure of $A_3^{II}B_2^{V}$ compounds in a large energy range (HUPFER, HIRSCH; HUPFER et al.).

In the present paper we report results of our efforts prior to the UPS and RELS measurements — the preparation of clean and stoichiometric surfaces of $A_3^{II}B_2^{V}$ crystals and thin films in an ultra-high vacuum (UHV) environment to achieve good reproducibility of surface conditions. Surface-sensitive (especially electron spectroscopic) measurements on $A^{II}B^V$ materials in the UHV were performed by only a few authors (Bishop et al.; Suda et al.; Domashevskaya and coworkers), in particular, surface conditions were not specified in detail so far. Proposals connected with etching procedures of {112} surfaces of $A_3^{II}B_2^V$ arsenides were published by Tajabor, Lovett. Contrary, Zivitz, Stevenson had obtained only limited success with etched samples and spoke in favour of only mechanically polished $A_3^{II}B_2^V$ surfaces.

Noble gas ion bombardment and subsequent surface reconstruction annealing are in their combination a widely used technique to prepare surfaces of different semiconducting compounds in the UHV. We have studied the influence of both methods on $A_3^{II}B_2^V$ surface chemistry whereas reconstruction (superstructures) or other structural phenomena were not investigated.

Thin stoichiometric $A_3^{II}B_2^V$ films with good structural and electric properties are usually prepared by means of simple direct evaporation (L. Żdanowicz, Miotkowska; L. Żdanowicz et al.; Pawlikowski, Becla). We have evaporated bulk material in the UHV to produce and characterize in situ deposited $A_3^{II}B_2^V$ thin films.

After describing sample (pre-UHV) treatment and experimental procedure in Sect. 2, the results of $A_3^{II}B_2^V$ UHV surface preparation monitored mainly by X-ray photoelectron spectroscopy (XPS) and Auger electron spectroscopy (AES) will be given in Sect. 3.

2. Experimental

2.1. Crystalline material

All $A_3^{II}B_2^V$ crystals studied were prepared by direct synthesis of stoichiometric quantities of highly purified elemental constituents As, P, Zn, and Cd. The material obtained was additionally purified from residual impurities (principally oxides) by manifold sublimations in evacuated closed ampoules. Details of crystal preparation and purification are described by Żdanowicz, Żdanowicz. Disc-shaped samples with specular natural faces, sufficient largeness (0.8 – 1.2 cm^2) and approximate 1 mm thickness were cut on a wire saw from single crystals in the case of Cd_3As_2 and Zn_3As_2. Analyzing crystal habits according to Kloc et al. the specimens were {112} oriented. However, crystallinity and orientation were not checked by diffraction methods. After cutting off from ingots polycrystalline disc-shaped samples of Cd_3P_2, Zn_3P_2 and $(Cd_{0.54}Zn_{0.46})_3As_2$ (hitherto $(Cd, Zn)_3As_2$) were mechanically polished with diamond paste and water-based alumina powder. Without furthermore polishing or etching procedure — only ultrasonically cleaning in pure acetone, trichlorethylen and ethanol — all metallic-shiny specimens were mounted on stainless steel sample holders, using a thin layer of Ag suspension to provide good electrical and thermal contact, and were immediately inserted into the UHV systems.

2.2. Deposited thin films

Following Bishop et al. (UPS/XPS study on dc-sputtered Cd_3As_2 and Zn_3As_2 thin films) some of our electron spectroscopic investigations were performed on $A_3^{II}B_2^V$ thin films. To avoid surface and bulk contamination, possibly from the first, Cd_3As_2 and Zn_3P_2 thin films were prepared by direct evaporation of bulk material in the UHV of the spectrometers. As source material small monocrystalline fragments were used. Since the studied $A_3^{II}B_2^V$ materials vaporize and condense congruently (Westmore et al.; Greenberg et al.), no special precautions must be taken to maintain stoichiometry of the vapour and the as-grown film.

The $A_3^{II}B_2^V$ starting material was contained in an acid-cleaned ceramic tube (∅ 2 mm) surrounded by a tungsten heater filament and enclosed at one end. The source-to-substrate distance was about 3 – 5 cm and a background pressure of about 10^{-6} Pa was maintained during the evaporation. The substrates were not moved into the vapour beam until the bulk material in the ceramic tube was well outgassed (examined by quadrupol mass spectrometer). The vapour beam was directed from the aperture of the tube to the substrate at near normal incidence (RELS/AES) or about 60° (UPS/XPS). For obtaining crystalline films (L. Żdanowicz, Miotkowska; L. Żdanowicz et al.) substrate temperatures were held at 475 K (Cd_3As_2) and 495 K (Zn_3P_2). Rates of deposition, thickness, and crystallinity of the prepared films were not determined. Stainless steel sample holders (UPS/XPS) and Si{111} samples (RELS/AES) (both cleaned by a sputter-anneal treatment) were used as substrate materials. After the fresh films were deposited, they were cooled down to room temperature and transferred without breaking the vacuum from the preparation into the analyzer chambers of the UHV devices as quickly as possible. All films prepared had a smooth uniform surface.

2.3. UHV devices and experimental procedure

The electron spectroscopic measurements described here were carried out in two separate UHV systems one equipped for photoemission and the other for RELS/AES studies.

The UPS/XPS experiments were performed in a commercial VG ESCA-3 device. The XP spectra were excited by nonmonochromatized Al, Mg Kα and Zr Mζ lines at 1486.6 eV, 1253.6 eV and 151.4 eV, respectively. The analyzer chamber ($p \leqq 5 \cdot 10^{-7}$ Pa) is attached to and isolated by a gate valve from the preparation chamber in which ion bombardment, deposition of thin films, and sample annealing took place. For Ar^+ ion bombardment (300/400 eV, 4—8 $\mu A/cm^2$) the preparation chamber was flooded by Ar gas ($p = 2 - 5 \times \times 10^{-3}$ Pa) which was purified by a LN_2 cooled sorption trap containing a molecular sieve. During ion bombardment and annealing a cold finger flanged in the sample vicinity was filled with LN_2. This procedure was carried out to suppress contamination of the whole UHV apparatus with deposited $A^{II}B^{V}$ particles and to minimize residual gaseous contamination that may outgas from the sample or the walls. Constant annealing temperatures of the specimens in the range of 430—530 K (measured by a thermocouple and caused by a heater inside the sample holder) were attained at about 3—5 min after establishing the heating current.

The RELS/AES data were taken on a home-built stainless steel UHV system (built up by AG Festkörperoberflächen, Sektion Physik) basically equipped with a 3grid retarding field analyzer (RFA). Inside the UHV recipient (background pressure of $3 \cdot 10^{-8}$ Pa) the samples were moved by a linear transport system into separate preparation or analyzer positions. An electron gun with a spotsize of nearly 1 mm², a variable beam energy up to 1.7 keV and normal incidence onto the sample was used in the AES/RELS experiments. AE spectra were taken with the usual first derivative ($dN(E)/dE$) mode and modulation amplitudes up to 10 V_{pp} applied onto the sample in order to suppress REL structures. Some of our AE spectra with low energy lines ($E_{XYZ} \leqq 200$ eV) were taken with a variable modulation voltage proportional to E or $E^{1/2}$ compensating the considerable decrease of $dN(E)/dE$ curves measured by RFA systems (Voigt). For Ar^+ ion bombardment (300 eV, 4—6 $\mu A/cm^2$, normal incidence) the whole vacuum chamber was flooded by research grade Ar gas ($p = 2 - 5 \cdot 10^{-3}$ Pa). Specimen annealing was realized using thermal radiation or additional electron bombardment from a tungsten filament onto the back of the samples. Substrate temperatures of about 470 K were easily attained (1—2 min) and monitored by a thermocouple situated near the specimen surface.

3. Results and discussion

3.1. Crystalline material

After the bakeout of the photoemission spectrometer (6 h, 425 K) XP spectra indicate that C, O, S, Cl, and F are the major contaminants on the $A_3^{II}B_2^{V}$ crystal surfaces (Figs. 1 and 2). In the vicinity of As and P core levels contributed by bulk atoms well

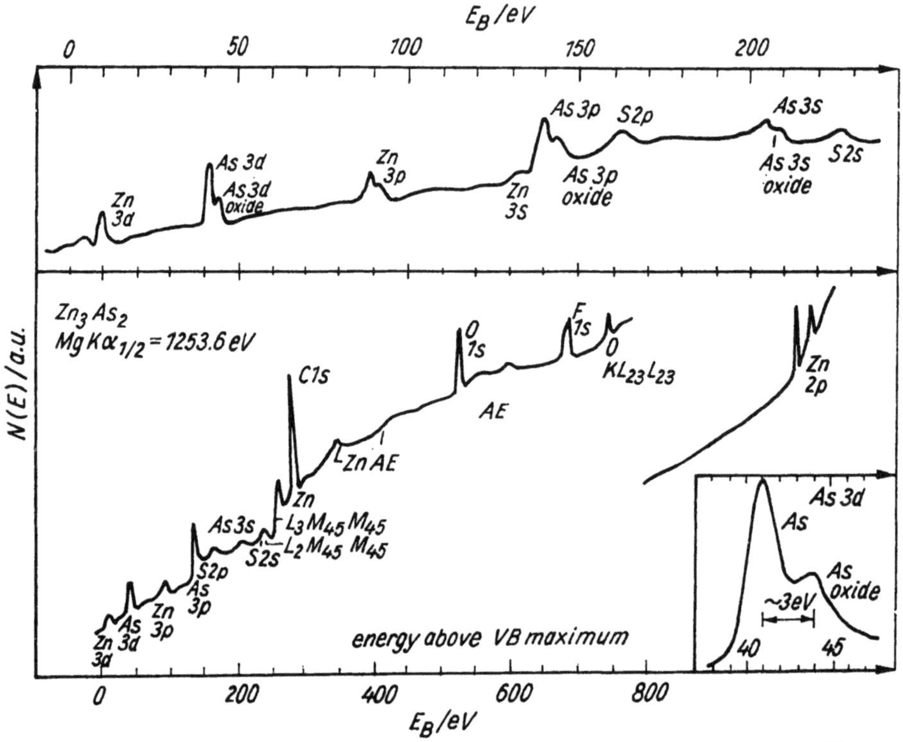

Fig. 1. XP spectra (Mg Kα excitation) of Zn_3As_2 after bakeout of the UHV device (index "oxide" labeled the As–O interaction) AE read below F

resolved additional chemical shifted peaks at higher binding energies than the main lines were observed. These peaks being characteristic of O–As/P interactions in the surface region are related to a charge transfer from the topmost anion atoms towards the stronger electronegative O atoms chemisorbed probably molecular from the residual gas onto the surface. The involvement of Cd and Zn surface atoms in this oxidation process is unclear because no detectable chemical shift in the core levels was observed although detailed fine structure and lineshape analysis was not carried out. In the case of $A_3^{II}B_2^V$ arsenides the O-induced chemical shift of the As 3d level was measured to be 3.0 eV (e.g.: Zn_3As_2 — Fig. 1 (insert)), for $A_3^{II}B_2^V$ phosphides the P 2p chemical shift was determined to be 4.5 eV (e. g.: Cd_3P_2 — Fig. 2 (insert)). The same order of magnitude of chemical shifts observed for $A_3^{II}B_2^V$ materials compared with GaAs (SU et al.) and InP (CHYE et al.) allows to predict similarities of $A_3^{II}B_2^V$ and $A^{III}B^V$ oxidation states.

After the bakeout of the RELS/AES spectrometer (2 × 10 h, 500 K) AE spectra indicated the same surface contaminants (C, O) and segregations (S, Cl) which were observed by XPS (Fig. 3a). Chemical shifted P LVV peaks at 108 eV and 95 eV were connected with H_2O/O-uptake from the residual gas discussed above and were absent in AE spectra of clean Cd_3P_2 or Zn_3P_2 surfaces. O-induced extra structures of P LVV around the energy values mentioned above are reported for oxidation states of $A^{III}B^V$ phosphides, too (WILLIAMS, MCGOVERN; MORGAN, VAN VELZEN), and were clearly detected in Zr Mζ excited XP spectra of Zn_3P_2 crystals (Fig. 3b).

In order to carry out a semiquantitative estimation of the elemental composition of $A_3^{II}B_3^V$ surfaces we have converted XPS peak heights as a measure for peak intensity

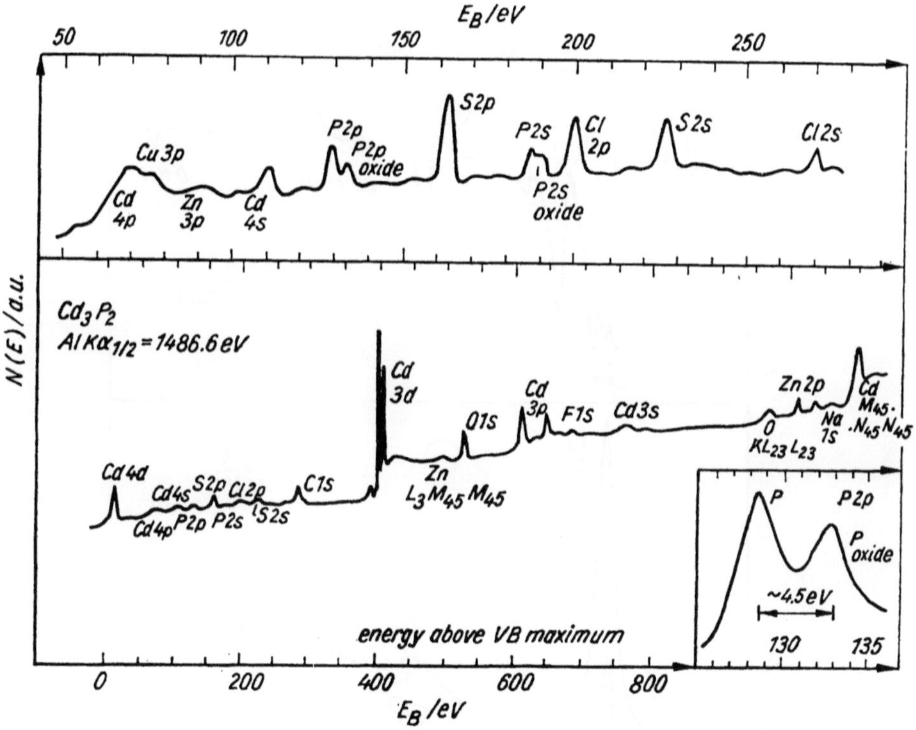

Fig. 2. XP spectra (Al Kα excitation) of Cd_3P_2 after bakeout of the UHV device (index "oxide" labeled the P–O interaction)

from Cd 4d, Zn 3d, As 3d, P 2p, C 1s, and O 1s lines into atomic fractions in the detected volume by using SCOFIELD's photoionization cross-sections and REILMAN's et al. values of asymmetric angular photoelectron distribution. The sum of all atomic fractions (A^{II}, B^{V}, C, O) is set to 100%. In the case of $(Cd, Zn)_3As_2$ solid solution Cd 4d and Zn 3d lines are superimposed, therefore, we have examined Zn 3p and Cd 3d peaks for quantitative analysis. Because of the considerable simplifications and approximations the error of our estimation may be probably high ($\lesssim$ 20%). Using AES, trends in surface composition caused by sample treatment were detected by monitoring APPH (Auger-peak-to-peak heights) ratios. For studied $A_3^{II}B_2^{V}$ crystals the Cd $M_{45}N_{45}N_{45}$ (379/386 eV), Zn $L_3M_{45}M_{45}$ (991 eV), As $L_3M_{45}M_{45}$ (1227 eV) and P $L_{23}VV$ (120 eV) Auger lines were recorded partly by using signal averaging and spectra accumulation (Zn and As LMM transitions).

The C- and O-covered surface layers have been removed by prolonged Ar^+ ion bombardment for 1–4 hours. The sputter treatment was interrupted for measuring XPS and AES depth compositional profiles. In Figure 4 we demonstrate a typical profile common for all $A_3^{II}B_2^{V}$ crystyls and both analytical tools studied. A deficiency of the B^{V} component caused by the high volatility of B^{V} oxides (As_4O_6, P_4O_{10}) under vacuum conditions was found in the outer surface layers. During sputter-cleaning A^{II}/B^{V} abundance ratios developed into a steady state, therefore, irrespective of further Ar^+ ion bombardment no compositional changes were observed by XPS or AES (Fig. 5 — curve "sputtering"). Agreement between A^{II}/B^{V} abundance ratios evaluated from XPS data and nominal 3:2 concentration ratio was found within experimental error. There-

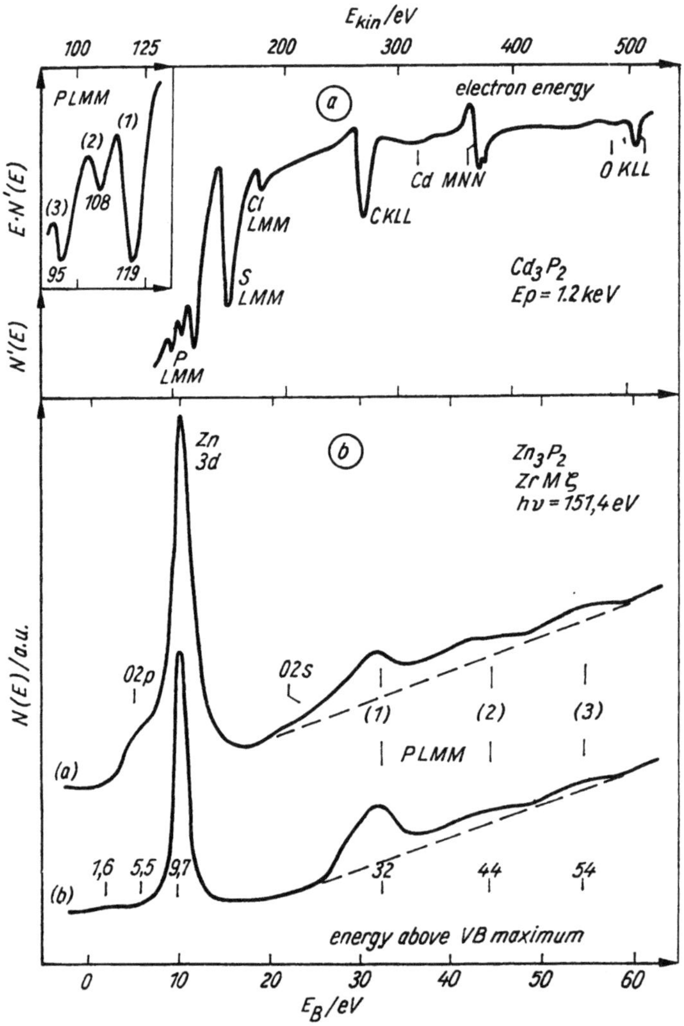

Fig. 3. a) Typical initial AE spectrum (E_p = 1.2 keV) of Cd_3P_2 crystalline material after bakeout of the UHV device (insert: O-influenced P LVV Auger transitions); b) Ultrasoft X-ray induced photoelectron spectra (Zr Mζ –151.4 eV) of Zn_3P_2 crystalline material with valence electron density of states structures and O-influenced P LVV Auger transitions for: (a) – a surface after bakeout procedure, (b) – a sputter-annealed surface in an early stage of preparation

fore, stoichiometric conditions seem to be reached using low energy Ar^+ ions and neither XPS nor AES depth profiles revealed preferential loss of one component of $A_3^{II}B_2^V$ surfaces. This result agrees with SUDA's et al. AES depth compositional profile on Zn_3P_2.

For an optimal recrystallization the annealing times and temperatures must be sufficient, the surface composition should be held stable, and annealing loss of one component should be suppressed. The following temperatures just enough for $A_3^{II}B_2^V$ surface recrystallization were determined by observing and minimizing intensity and noise of secondary, scattered electrons in UP and REL spectra versus temperature: Cd_3P_2 – 430 K, Cd_3As_2 and Zn_3As_2 – 450 K, (Cd, Zn)$_3$As$_2$ – 460 K and Zn_3P_2 –

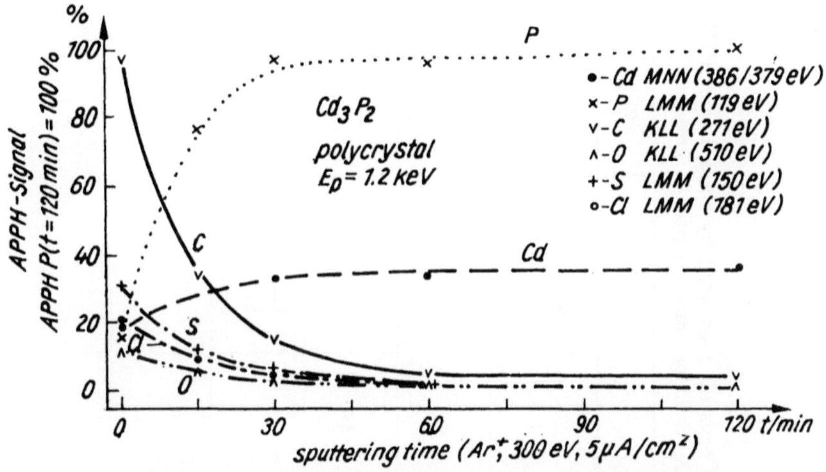

Fig. 4. Typical AES depth-compositional profile of $A_3^{II}B_2^{V}$ crystals (e.g.: Cd_3P_2)

480 K. By raising annealing temperatures, enrichment of the B^V component in the surface region was observed above 520 K (Cd_3P_2: 500 K) and/or, only with a slight tendency, by using annealing times longer than 30 min at temperatures mentioned above. In particular, P excess in phosphides seems to be stronger than As excess in $A_3^{II}B_2^{V}$ arsenides analyzing A^{II}/B^{V} XPS or AES intensity ratios in detail (Fig. 5 — curve "annealing": Cd_3P_2 as the prominent example of B^V excess). Prolonged heating (> 1 h) leads, besides low B^V excess, to a reappearance of S and Cl traces, presumably due to diffusion from the bulk onto the surface.

After the prolonged Ar^+ ion bombardment (1—4 h) and repeated (2—5) preparation cycles of Ar^+ ion bombardment (15—30 min) and annealing (15—20 min) the surface

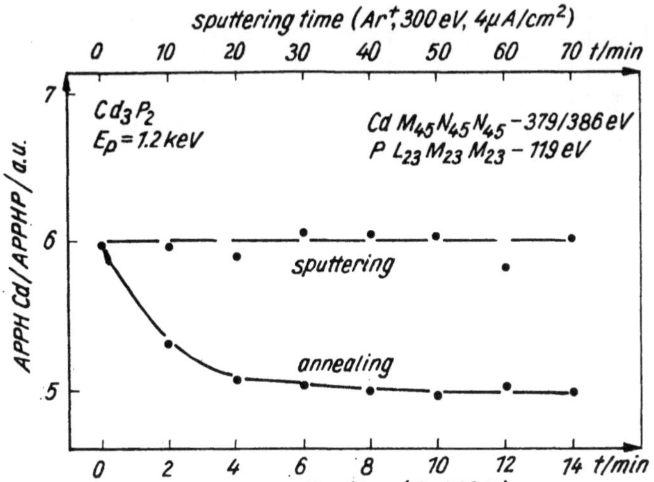

Fig. 5. Auger-peak-to-peak heights (APPH) ratios of A^{II}/B^{V} elements (e.g.: Cd_3P_2) versus sputtering and annealing times

Table 1

Comparison between elements weight concentrations (in mg · g^{-1}) of a) $A_3^{II}B_2^V$ crystals measured by thick target PIXE (information depth ~ 5 μm), b) sputter-annealed $A_3^{II}B_2^V$ crystals measured by XPS (information depth ~ 5 nm)

		Cd_3As_2	$(Cd_{0.54}Zn_{0.46})_3As_2$	Zn_3As_2	Cd_3P_2	Zn_3P_2
Cd	stoichiometric	692	431	—	845	—
	PIXE	688 ± 15	424 ± 10	3.5 ± 1.0	852 ± 15	—
	XPS	695	367	—	843	—
Zn	stoichiometric	—	214	567	—	760
	PIXE	—	212 ± 10	574 ± 15	—	775 ± 15
	XPS	—	256	583	—	792
As	stoichiometric	308	355	433	—	—
	PIXE	292 ± 10	331 ± 10	411 ± 10	—	—
	XPS	269	363	403	—	—
P	stoichiometric	—	—	—	155	240
	PIXE	—	—	—	148 ± 20	194 ± 25
	XPS	—	—	—	143	188
other elements	PIXE	20	33	11.5	0	31
C	XPS	18	8	9	9	5
O	XPS	18	6	5	4	15

contamination was removed close to the XPS and AES detection limits. In Table 1 elemental weight concentrations of these ready prepared $A_3^{II}B_2^{V}$ crystal surfaces evaluated by surface-sensitive XPS are collected and compared (without discussion) with bulk concentrations estimated by particle-induced X-ray emission spectroscopy (PIXE)[1]).

Stoichiometric conditions checked by XPS were monitored in the case of the $(Cd, Zn)_3As_2$ solid solution (nominally composed of 54% Cd and 46% Zn for A^{II}) subjected to prolonged Ar^+ ion bombardment. Further sputter-anneal treatment (identical to other $A_3^{II}B_2^{V}$ crystals in the UHV) caused a low Zn enrichment in comparison to Cd (45% Cd and 55% Zn for A^{II}) while the A^{II}/B^{V} abundance ratio was ideal and constant with respect to the different preparational steps. Similar stoichiometric changes characterized by a preferential loss of the element standing above in the II^B column of the periodic system were observed during UHV preparation (gentle ion bombardment, annealing) of $Hg_{1-x}Cd_xTe$, where Cd tends to compensate Hg loss, partially (NITZ et al.; WEBB et al.).

The sticking probabilities of residual gases on the freshly prepared $A_3^{II}B_2^{V}$ surfaces are small in ambient temperature, if electron beam effects were turned off.

3.2. Thin films

Immediately after preparing evaporated thin films, XP spectra of Cd_3As_2 and Zn_3P_2 were recorded in situ. No C or O contamination could be detected by checking the O 1s, C 1s, and O KLL signals in the XP spectra (Fig. 6). Practically no difference between A^{II}/B^{V} abundance ratios of in situ deposited thin films and ion bombarded (and subsequent annealed) crystalline material was obtained. Taking into consideration that the in situ prepared thin film surface was not handled by any treatment (especially no ion bombardment) this result supports the consistency of our measurements: Low energy Ar^+ ion bombardment applied to $A_3^{II}B_2^{V}$ crystal surfaces leads not only to steady molar composition but also to really stoichiometric conditions. By analogy with investigations on crystals XPS depth profiles of in situ prepared films were carried out by ion sputtering. No traces of contamination or compositional changes were monitored.

In comparison to in situ prepared thin films some Cd_3As_2 films evaporated in common oil-diffusion pumped vacua (10^{-3} Pa, 470–480 K temperature of mica substrates) were investigated. A XPS depth-compositional profile illustrates (Fig. 7) that C and O atoms were not only concentrated within a thin surface layer caused by transport on air, but are incorporated throughout the film. The significant contribution of the C 1s XPS line remains unchanged by additional prolonged ion bombardment. Within experimental error stoichiometric composition of these C-rich Cd_3As_2 thin films was found by examining XP spectra.

4. Summary/Conclusion

Different $A_3^{II}B_2^{V}$ surfaces were prepared under UHV conditions in order to carry out density of states investigations (UPS, RELS). Cd_3As_2 and Zn_3P_2 thin films characterized by the lack of contamination were deposited from bulk material onto steel or silicon substrates. Clean and stoichiometric surfaces of $A_3^{II}B_2^{V}$ crystalline material were accomplished by cycles of a sputter-anneal treatment. In the limit of experimental error the A^{II}/B^{V} abundance ratios of crystalline material (prepared by sputter-annealing) and thin films (UHV deposited and in situ analyzed) were in excellent agreement

[1]) The authors thank Dr. H. FREY from WB Angewandte Kernphysik (Prof. Dr. G. Otto), Sektion Physik, KMU Leipzig for these studies.

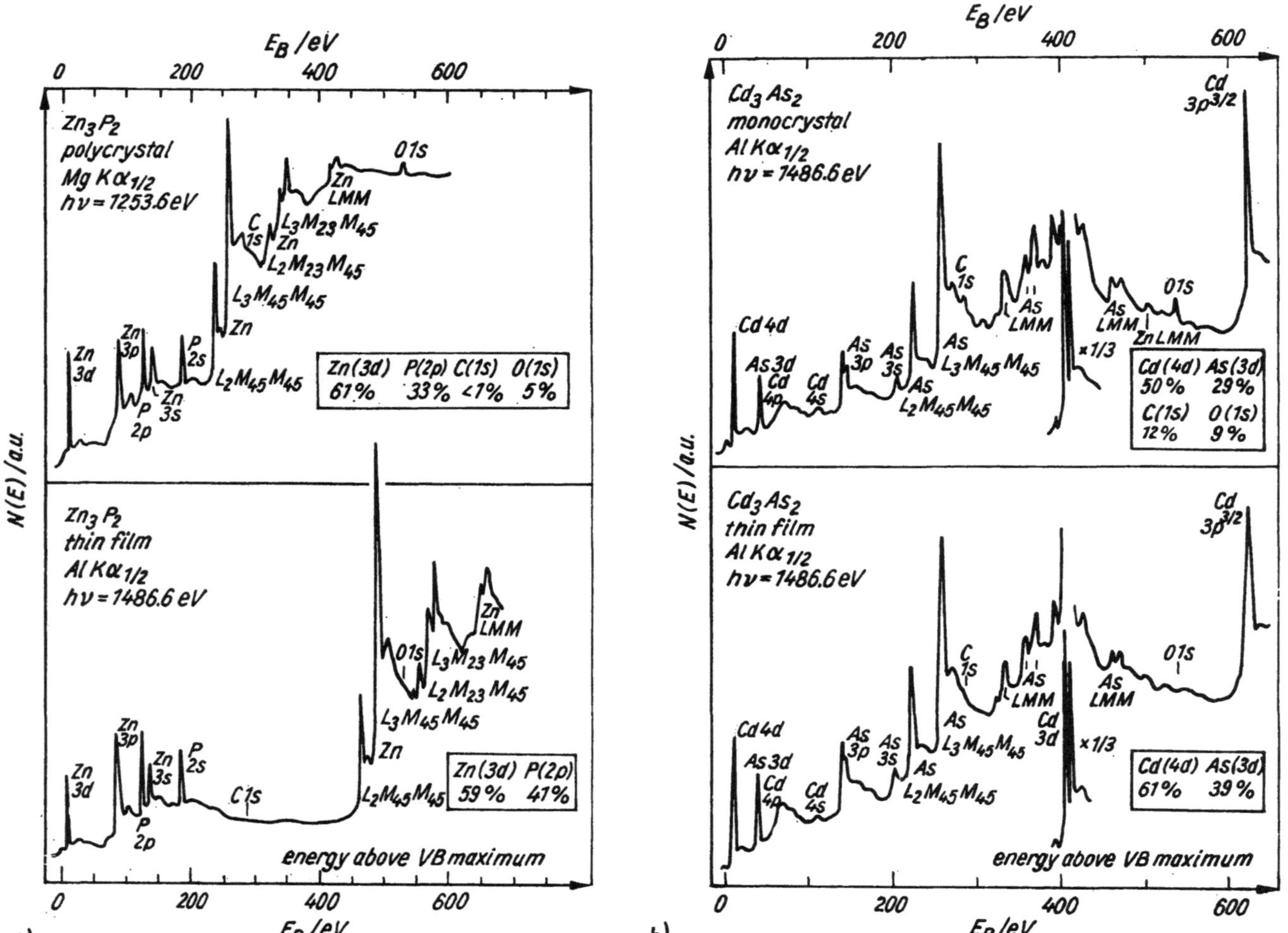

Fig. 6. a) Comparison between XP spectra of a well prepared polycrystalline surface and a fresh UHV deposited thin film of Zn_3P_2. Different Auger lines positions are caused by different used X-ray excitation energies (inserts: estimated atomic fractions); b) Comparison between XP spectra of a sputter-annealed monocrystalline surface and a fresh UHV deposited thin film of Cd_3As_2 (inserts: estimated atomic fractions)

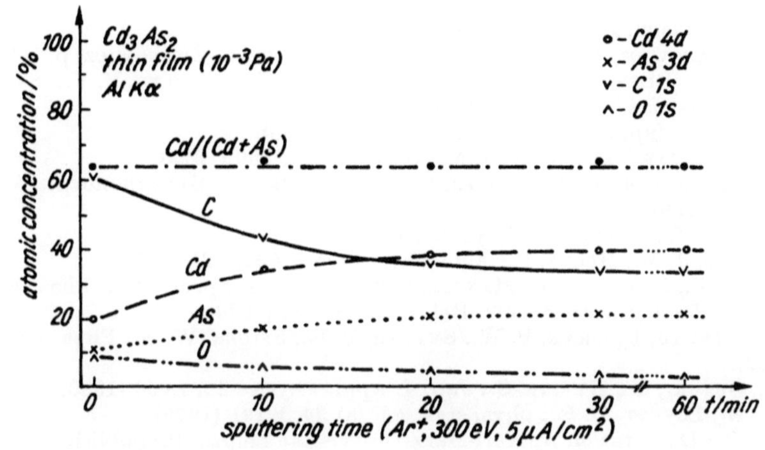

Fig. 7. Typical XPS depth-compositional profile of a Cd_3As_2 thin film deposited in usual oil-diffusion pumped vacuum of 10^{-3} Pa

with each other and the nominally $A_3^{II}B_2^{V}$ composition. A preferential sputtering of one component was not observed, a B^V enrichment occurs by raising the annealing temperatures above 520 K.

Generally spoken: Working with $A_3^{II}B_2^{V}$ materials under UHV conditions is possible and similar to procedures reported in the case of other semiconducting compounds.

A. H., S. Sch. and D. H. thank Prof. Dr. W. Żdanowicz and Doz. Dr. Ch. Kleint for leading us to this subject, for help and discussion. Two of us (A. H. and S. Sch.) thank WB Physikalische Chemie (Prof. Dr. A. Meisel) of Karl-Marx-Universität for the possibility of performing the UPS/XPS measurements in the VG ESCA-3 spectrometer. We wish to acknowledge the skillful technical assistence of our colleagues Dr. D. Flamm, Dr. W. John, Dr. A. Reif and DC A. Hofmann during the photoemission experiments. Representative for other coworkers, supporting this work, we mention Dr. K. G. Voigt for designing and constructing the AES/RELS spectrometer electronics.

References

Bishop, S. G., Shevchik, N. J., Cardona, M.: in Proc. IV. Intern. Conf. Vac. UV Rad. Phys., Oxford, Braunschweig 1976, p. 497

Chye, P. W., Su, C. Y., Lindau, I., Garner, C. M., Pianetta, P., Spicer, W. E.: Surf. Sci. 88, 439 (1979)

Domashevskaya, E. P., Terekhov, V. A.: a) in: Proc. X84 Intern. Conf. X-Ray and Inner-Shell Proc. in Atoms, Molec. and Solids, Leipzig 1985, p. 357, b) Izv. Akad. Nauk SSSR, Ser. Fiz. **49**, 1562 (1985)

Domashevskaya, E. P., Ugai, J. A.: Izv. Sibir. Otd. Akad. Nauk SSSR, Ser. Khim. Nauk **4**, 49 (1975)

Frey, H., Vogt, J., Otto, G.: J. Radioanal. Nucl. Chem. **99**, 193 (1986)

Gelten, M. J., Żdanowicz, L. (eds): Proc. First Intern. Symp. Phys. Chem. II—V Comp., Eindhoven 1981

Greenberg, J. H., Lazarev, V. B., Kozlov, S. E., Shevchenko, V. J.: J. Chem. Thermodyn. **6**, 1005 (1974)

Hupfer, A., Hirsch, D.: a) Acta Univ. Wratislav. **936**, 153 (1986), b) Physik d. Halbleiteroberfläche **15**, 15, 23 (1984), Berlin 1984

Hupfer, A., Hirsch, D., Schulze, S.: to be published
Kloc, K., Żdanowicz, W., Cisowska, E.: ibid. Gelten, L. Żdanowicz, p. 109
Lazarev, V. B., Shevchenko, V. J., Greenberg, J. H., Sobolev, V. V.: Semiconductor Compounds of $A^{II}B^{V}$ Group, Moscow 1978 (in Russian)
Matsumoto, Y.: Jap. J. Appl. Physics **22**, 1699 (1983)
Morgan, A. E., van Velzen, W. J. M.: Surf. Sci. **40**, 360 (1973)
Nitz, H. M., Ganschow, O., Kaiser, U., Wiedmann, L., Benninghoven, A.: Surf. Sci. **104**, 365 (1981)
Pawlikowski, J. M.: Infrared Phys. **21**, 181 (1981)
Pawlikowski, J. M., Becla, P.: Acta Phys. Polon. **A47**, 121 (1975)
Reilman, R. F., Msezane, A., Manson, S. T.: J. Electr. Spectr. Rel. Phen. **8**, 389 (1976)
Scofield, J. H.: J. Electr. Spectr. Rel. Phen. **8**, 129 (1976)
Su, C. Y., Lindau, I., Chye, P. W., Skeath, P. R., Spicer, W. E.: Phys. Rev. **B25**, 4045 (1982)
Suda, T., Suzuki, M., Kurita, S.: Jap. J. Appl. Physics **22**, L656 (1983)
Tajabor, N., Lovett, D. R.: phys. stat. sol. (a) **34**, K191 (1976)
Voigt, K. G.: Dissertation A, Karl-Marx-Universität Leipzig (unpublished)
Webb, C., Lagowski, J., Lichtensteiger, M.: Surf. Sci. **138**, 399 (1984)
Westmore, J. B., Mann, K. H., Tickner, A. W.: J. Phys. Chem. **68**, 606 (1964)
Williams, R. H., McGovern, I. T.: Surf. Sci. **51**, 14 (1975)
Żdanowicz, L., Miotkowska, S.: a) Thin Solid Films **29**, 177 (1975), b) J. Cryst. Growth **48**, 387 (1980)
Żdanowicz, L., Żdanowicz, W., Petelenz, D., Kloc, K.: Acta Phys. Polon. **A57**, 159 (1980)
Żdanowicz, W.: II_x—V_y Compounds, in: Landolt-Börnstein, Numerical Data and Functional Relationships in Science and Technology, New Series, Group III, Vol. 17/e, Chap. 9.5., Berlin, Heidelberg, New York, Tokyo 1983, p. 178
Żdanowicz, W., Żdanowicz, L.: Ann. Rev. Material Sci. **5**, 301 (1975); Ann Rev. Inc., ed. Huggins, Palo Alto (Calif.) 1975
Zivitz, M., Stevenson, J. R.: Phys. Rev. **B10**, 2457 (1974)

(Received 1) December 16, 1986;
2) January 28, 1987)

Authors' addresses:

Dr. A. Hupfer
Karl-Marx-Universität Leipzig, Sektion Physik, Wissenschaftsbereich Halbleiterphysik
Linnéstraße 5
Leipzig
DDR-7010

Dr. S. Schulze (present address:)
Technische Universität Karl-Marx-Stadt,
Sektion Physik/Elektronische Bauelemente
Wissenschaftsbereich Struktur der Materie
PSF 964
Karl-Marx-Stadt
DDR-9010

Dipl.-Chem. D. Hirsch (permanent address)
Karl-Marx-Universität Leipzig
Sektion Chemie, Wissenschaftsbereich Physikalische Chemie
Talstraße 35
Leipzig
DDR-7010

Doc. Dr. L. Zdanowicz
Polska Akademia Nauk, Zakład Fizyki Ciała Stałego
ul. Jana Kawalca 3
41800 Zabrze, Poland

Cryst. Res. Technol.	22	1987	7	923–928

P. SEDLACEK, R.-J. KUBAN, K.-O. BACKHAUS

Zentralinstitut für physikalische Chemie
der Akademie der Wissenschaften der DDR, Berlin

Structure Determination of Polytypes (II)

In this part different methods for structure determination of polytypes with diffraction methods are described. These methods are applied to periodic polytypes consisting of one or more sorts of layers and to disordered polytypes consisting of one sort of layers. Furthermore an example for the determination of the structural principle is given.

Verschiedene Methoden zur Strukturbestimmung von Polytypen werden in diesem Teil behandelt. Die angeführten Methoden wurden für periodische Polytypen, die aus einer oder mehreren Sorten von Schichten, sowie für fehlgeordnete Polytypen, die aus einer Sorte von Schichten bestehen, genutzt. Ferner wird ein Beispiel zur Bestimmung des Strukturprinzips angegeben.

2.1. Patterson methods

2.1.1. Periodic polytypes consisting of one sort of layers

Amongst the polytypes consisting of one sort layers are those with a structure corresponding to close packed spheres, as SiC, ZnS, CdS etc.

A number of authors deals with Patterson methods for the determination of layer stacking sequences of long-periodic polytypes, as TOKONAMI; DANIELS; DORNBERGER-SCHIFF and FARKAS-JAHNKE; FARKAS-JAHNKE; FARKAS-JAHNKE and DORNBERGER-SCHIFF; GOMES DE MESQUITA; KAKINOKI et al. 1969, 1972; FARKAS-JAHNKE and GACS. All these methods are based on the determination of the $N_{m,n,p}$-values with Patterson functions and their deconvolution. DORNBERGER-SCHIFF and FARKAS-JAHNKE have given a very obvious method which will be compared with other methods. The basic relations related to an orthohexagonal coordinate system are

$$K_{(k,l)} = \frac{I_{(k,l)}}{|F^{(0)}_{(k,l)}|^2},$$

$$K_{(k,l)} \sim |S_{(k,l)}|^2\,, \tag{I}$$

$$K_{(k,l)} \sim \sum_{n=0}^{2} \sum_{p=0}^{Q-1} N_{n,p} \exp\left[2\pi i\left(\frac{kn}{3} + \frac{lp}{Q}\right)\right], \tag{II}$$

with $N_{0,0} = Q = N$ (hexagonal and trigonal)
$N_{0,0} = 3N = Q$ (rhombohedral).

The corresponding π-function is given by

$$\pi_{(n,p)} = \frac{N}{3}\left[1 + 2\sum_{l=0}^{Q-1} \frac{|S_{(1,l)}|^2}{Q^2} \cos 2\pi\left(\frac{n}{3} + \frac{lp}{Q}\right)\right], \tag{III}$$

with

$$\sum_{l=0}^{Q-1} |S_{(1,l)}|^2 = Q^2 .$$

The layer sequences γ_p of the length p is represented by digits 0(↙) and 1(↘), e.g. $\gamma_p = (00101 \ldots)$. The frequencies of the sequences is denoted by $[\gamma_p]$. Transforming the $\pi_{n,p}$-function into a function $\psi(K, p)$ using the condition $m + p + K \equiv 0 \pmod 3$ we obtain the relations between the stacking sequences and the Patterson function. K is the total of digits of a number (mod 3) of a sequence of digits with the length p, e.g. $[000] + [111] = \psi(0, 3)$. Further relations using the properties of the π-function are given by

$$[00] = [000] + [001] = [000] + [100]$$

or in general

$$[\gamma_p] = [\gamma_{p0}] + [\gamma_{p1}] = [0_{\gamma p}] + [1_{\gamma p}] .$$

These relations are successively used for the deconvolution of the Patterson function.

At first TOKONAMI has used a one-dimensional Patterson function $V(z)$ for the determination of a 96R-polytype of SiC. The Patterson function $V(z)$ gives the same informations as the π-function in rhombohedral cases. In contrast to FARKAS-JAHNKE (1966) and FARKAS-JAHNKE and DORNBERGER-SCHIFF (1970), applying the π-function and used very accurate intensity values, TOKONAMI determined the $|S|^2$ values with lesser accuracy and corrected them by necessary conditions for the $|S|^2$ and the $N_{m,n,p}$-values. GOMES DE MESQUITA determined the structure of a 120R polytype of SiC. KAKINOKI et al. (1969, 1972) used two one-dimensional Patterson functions C_p and S_p, which give the same information as the two-dimensional π-function.

2.1.2. Disordered polytypes consisting of one sort of layers (pseudo-lattice structures)

From (VI) of Part I follows

$$1/PK = 1/P \sum_{m=0}^{Q_1-1} \sum_{n=0}^{Q_2-1} \sum_{p=-(P-1)}^{(P-1)} N_{m,n,p} \exp\left[2\pi i(hm/Q_1 + kn/Q_2 + \xi p)\right] \quad \text{(IV)}$$

This equation may be compared directly with the analogous representation of KAKINOKI and KOMURA (1965, Part I)

$$1/PK = 1 + \sum_{p=1}^{P-1} \left[(1 - p/P) \exp\left(-2\pi i p\xi\right)\right] T_p(h, k) + \text{conj.} \quad \text{(V)}$$

with $T_p(h, k) = \text{spur}\, \varepsilon F P^p$. The normalized function therefore gives as shown by KAKINOKI and KOMURA the probabilities $N'_{m,n,p} = (N_{m,n,p}/P)$, for the displacement of two layers by the vector (m, n, p).

The deconvolution of the function applied to periodic polytypes by analogous methods leads to the determination of relative frequencies $[\gamma_p]/P$ of the layer stacking sequences. A direct method for the determination of the $N_{m,n,p}$-values has been published by FARKAS-JAHNKE (1966) and used for the determination of $[\gamma_p]/P$ $(p = 4)$ ZnS crystals (FARKAS-JAHNKE, GACS). The diffuse rods for $k = 1$ (orthohexagonal system have been measured at points $\xi = l/Q_3$ (with l as integers). Thus, the structure has an artificial period $(\boldsymbol{c} = Q_3\hat{\boldsymbol{c}})$.

The $N'_{m,n,p}$-values obtained by Fourier transformation of the K-values are now superpositions of the real $N'_{m,n,p}$-values on the artificial period, and are thus falsified. This falsification will be reduced by the kind of measurement of the intensities with finite slit or otherwise, due to the gradual equalizing of $N_{0,p}$, $N_{1,p}$, and $N_{1,p}$ for grate p.

In disordered cases FARKAS-JAHNKE (1966) used intensities, measured at points $\xi = l/Q_3$ with different Q_3 values and comparing the changes of the $N'_{m,n}$-values as a degree of falsification.

2.1.3. Polytypes consisting of two or more sorts of layers

Considering at first disordered polytypes, we start from relation (IV) in Part I. The meaning of the $K^{(s,t)}$-function is given by the equation (VI) in Part I, if the structures are pseudo-lattice-like. By Fourier transformation of the $K^{(s,t)}$-functions we obtain the $N^{(s,t)}_{m,n,p}$-values, stating the probabilities, that two layers (one of sort s and one of sort t) are displaced by the vector (m, n, p). The corresponding π-function may be deconvoluted by methods analogous to those described in the former part of this paper. There is the problem, to separate the $K^{(s,t)}$-function from the intensity contributions of the layers $[F(\tilde{s})\ F(t)]$. The problem remains unsolved, if the more general relations formulated by KAKINOKI and KOMURA (1965, I) are used. They are obtained if in (XIV, Part I) T_p is substituted by $I_p =$ spur VFP^p with $(V)_{ji} = F_i\tilde{F}_j$. For pseudo-lattice polytypes the periodicity of the $K^{(s,t)}$-functions may be used, because these functions remain unchanged within a reflection class (λ), whereas the contributions of the layers are changed. If linear dependences are not present, the $K^{(s,t)}$-functions may be calculated, for instance by least quares methods. For two sorts or kinds of layers such methods (SEDLACEK 1978, Part I; WALTHER) and the corresponding computer programs (KRETSCHMAR and WALTHER, unpublished) have been worked out. These methods have already been used for the refinement of some structure principles.

A special Patterson method for two kinds of layers has been used by KIHARA and MATSUMOTO (1978) for the determination of the stacking sequences of the substances BiTe, $Bi_4(S, Se)_3$, and Bi_8Se_9. Only the reflections (001) have been taken into account:

$$F_{(00l)} = F_l = S_l^{(0)}F_l^{(0)} + S_l^{(1)}F_l^{(1)}\,, \tag{VI}$$

$$S_l^{(s)} = \sum_{p=0}^{Q-1} g_p^{(s)} \exp\left(2\pi \frac{lp}{Q}\right),$$

$$g_p^{(0)} + g_p^{(1)} = 1\,,$$

$$F_{(00l)} = S_l^{(0)}(F_l^{(0)} - F_l^{(1)}) + Q \cdot F_l^{(1)}\delta(l \to 0)\,.$$

With these relations the $K^{(s,s)} = |S^{(s)}|^2$ function was determined. The corresponding one-dimensional π-function gives the sequence of the kinds of layers.

3. Determination of the structure principle

In many cases it is possible to determine the structure principle without knowledge of the total intensity distribution of the general diffuse reflections. In these cases the intensity values have to be measured at characteristic points of the reciprocal space, with the aid of the peak method. Essentially we obtain information on the structure of single layers, but not specific statements on the stacking of the layers. For determination of pseudo-lattice polytypes it is useful to utilize single reflection classes (λ). Analogous classes may be used also for two- or one-dimensional pseudo-lattice polytypes, where the periodicity of the $S^{(s)}$ and $K^{(s,t)}$ functions are two- or one-dimensional. The corresponding Patterson function or electron densities, belonging to one reflection class (λ), are fictitious, but their interpretation gives information on the structure principle (SEDLACEK 1978; SEDLACEK, DORNBERGER-SCHIFF 1965, 1971; KUBAN, SEDLACEK).

The most simple case is the structure determination with one sort of layers. The intensities are of the form

$$I = K^{(0,0)}F^{(0)^2} = K^{(0,0)} \cdot \tilde{F}^{(0)} \cdot F^{(0)} . \quad \text{(VII)}$$

Therefore the $K^{(0,0)}$ functions, disregarding their meaning, may be used as constants within any reflection class or partial reflection class (λ).

In many cases it is possible to combine certain classes (λ) of reflections in such a way, that the obtained intensity set (*hkl*) corresponds to fictitious but periodic structures. To obtain such a set (*hkl*) the corresponding single classes of reflections must be scaled. Using certain classes of intensity maxima, periodic structures can be constructed, which correspond to ordered domains of the disordered structures. For such a scaled set of intensities the usual methods of structure determination for periodic substances may be applied. (SEDLACEK 1978; KUBAN, SEDLACEK).

If the structure consists of more than one sort of layers, some complications appear. In general the intensity values cannot be transformed into structure factor values (see also equation (III).

In some cases it is possible to obtain the structure factor values and thus electron density functions by transformations of intensity values. The pseudo-lattice structures consisting of more than one sort of layers may be refined by least squares methods as mentioned above. The refinement adjust not only the structure parameters but also the $K^{(s,t)}$ values, which belong to the different reflection classes (λ) of a certain set of intensities (*hkl*).

For application of such methods preliminary information or starting hypothesis about the building principles of the structure are necessary. This preliminary information follows from the chemistry of the substances, from a comparison of structures of the same polytype class or the same OD family.

In the following we will show an example, how we can obtain such preliminary information only from the diffraction pattern with the assumption of an OD structure. The diffraction pattern from the modification of dinitrosylcobalt chloride shows apparent sharp reflections but unusual extinctions, which may be interpreted, if we assume the following classification ($Q = 2$ and $h_\lambda = 0$ and 1) (KUBAN, SEDLACEK). Therefore two classes of reflections result with $h = 2\hat{h}$ and $h = 2\hat{h} + 1$. The following systematic conditions for the two classes were noted:

$h = 2\hat{h} + 0$		$h = 2\hat{h} + 1$	
$\hat{h}kl$:	$\hat{h} + l = 2n$	hkl:	no conditions
$\hat{h}0l$:	$\hat{h} + l = 2n$	$h0l$:	absent
$0kl$:	$k = 2n$		

From conditions for $h = \hat{2h}$ it follows that the space group of superposition structure (40) is Bb. and from conditions for $h = 2\hat{h} + 1$ follows that the layer group is P.a(.). From the a-glide planes of layers and from $\hat{\boldsymbol{a}} = \boldsymbol{a}/2$ it follows that the superposition structure has mirror planes perpendicular to $\boldsymbol{b}$. Thus Bbmm or Bbm2 result as the probable space groups of the superposition structure. The minimum symmetry of the single layer is Pba(2), and Pba(m) is also consistent with the space groups of the superposition structure and the absences. The result of investigations of physical properties have led to the conclusion that the structure is centrosymmetrical. Thus we may assume the symmetry of single layer is Pba(m). The different layers of the OD structure are translationally equivalent and adjacent layers are related by a vector $\boldsymbol{a}/2 + \boldsymbol{c}_0$ or $-\boldsymbol{a}/2 + \boldsymbol{c}_0$, where $\boldsymbol{c}_0$ is the distance between adjacent layers. The result of the symmetry of the single layer and these translations is the OD groupoid family:

$$\begin{matrix} \mathrm{P} & \mathrm{b} & \mathrm{a} & (\mathrm{m}) \\ \{n_{1,2} & n_{2,\bar{1}/2} & (a_{1/2})\} \end{matrix}$$

Thus the structure is pseudo-lattice-like and with this hypothesis the structure determination will start. Only two reflection classes $h_\lambda = 0$ and $h_\lambda = 1$ are observed. The first corresponds to a fictitious structure, the socalled superposition structure, e.g. these reflections are sharp. The other corresponds to a superposition of the form $1/2[\varrho(x, y, z) - \varrho(x + 1/2, y, z)]$. These reflections are in general diffuse. In this special case only apparent sharp reflections are observed and nevertheless disorder must be assumed as working hypothesis, e.g. these reflections are in reality maxima of intensities. The reflections of both classes were measured by peak method and scaled by a factor proportional to $1/K_\lambda$. The obtained set of reflections (hkl) corresponds in this case to one of the two possible MDO structures of the OD family. With such methods the structure principle of a number of structures was determined (SEDLACEK, DORNBERGER-SCHIFF 1965, 1971; JAGNER et al. 1974, 1977).

An analogous method as described above for determination of the structure principle of structures consisting of one sort of layers is given by MÜLLER using the intensity relation derived by KAKINOKI ynd KOMURA (1965, I)

$$I = F_0 \cdot \tilde{F_0}[ND(\xi) + H(\xi)] . \tag{VIII}$$

MÜLLER considered the special case where the reflections with $h = 2\hat{h}$ are sharp and the reflections with $h = 2\hat{h} + 1$ are diffuse rods in $\boldsymbol{c}^*$ direction without maxima. Supposing the knowledge of the superposition structure corresponding to reflections $h = 2\hat{h}$ with the fictitious period $\boldsymbol{a} = 2\hat{\boldsymbol{a}}$, $\boldsymbol{b} = \hat{\boldsymbol{b}}$, $\boldsymbol{c} = \hat{\boldsymbol{c}}$, he derives models for the structure of a single layer, which corresponds to the superposition structure. For these models the intensities, given by the relations above, are calculated for certain sets of intensities with $\xi =$ const. and compared with the observed intensities. The problem is the measurement of the intensities, if diffuse rods without intensity maxima are observed.

An example of a disordered structure consisting of four sorts of layers is the structure of vaterite, μ-$CaCO_3$, analysed by MEYER using methods of JAGODZINSKI (cited in Part I). A Patterson function calculated with the observed strong maxima $(h, h - 3n, l)$ only gives preliminary parameters about the structure principle. The intensity distribution was calculated for strong maxima with $h - k = 3n$ as

$$I \sim [|F_1 + F_{\bar{1}}|^2 + \cos 2\pi\xi \cdot Re(F_1 + F_{\bar{1}})^2] \frac{\sin^2 \pi\xi P}{2 \sin^2 \pi\xi} + I_d . \tag{IX}$$

Assuming "Reichweite" $s = 2$ for the reflections with $h - k \neq 3n$ the intensity relation was derived as

$$I \sim \frac{P}{2} (|F_1|^2 + |F_{\bar{1}}|^2) \cdot f(\xi) . \tag{X}$$

With the help of the first intensity relation (IX) the superposition structure was refined using least squares methods. Two structure models were refined by least squares methods using reflections with $h = k \neq 3n$ and the corresponding intensity relation (X), but it was not possible to distinguish between both structure models.

Better results may be obtained by application of methods as described above, which use the lattice properties of this structure. Thus the following relations are given

$$F = \sum_{s=0}^{3} S^{(s)} F^{(s)} , \tag{XI}$$

i.e. the structure is composed of four sorts of layers and

$$S^{(s)} = \sum_{m=0}^{2} \sum_{p=0}^{P-1} g_{m,p}^{(s)} \exp\left[2\pi i \left(\frac{km}{3} + \xi p\right)\right] \quad \text{(XII)}$$

i.e. the $S^{(s)}$ functions are periodic, therefore they are constant within a whole class of reflections (λ) e.g. $k = 3\hat{k} + k_\lambda$, $\xi = \xi_\lambda + \hat{l}\xi$.

Using the meaning of the Fourier transformation of a single class of reflections a structure model may be derived and refined by special least squares programs as described above, varying not only structure parameters of single layers but also the $K^{(s,t)}$ functions which are constant within a whole class of reflections (λ).

Using the symmetry relations as given by MEYER

$$F^{(0)} = \tilde{F}^{(1)} = Re_1 + iIm_1$$

and

$$F^{(1)} = \tilde{F}^{(0)} = Re_{\bar{1}} + iIm_{\bar{1}} \quad \text{(XIII)}$$

the structure factor relation is obtained as follows

$$S=(S^{(0)}+S^{(2)})\, Re_1+(S^{(1)}+S^{(3)}Re_{\bar{1}}+i(S^{(0)}-S^{(2)}Im_1+(S^{(1)}-S^{(3)})\, Im_{\bar{1}}\,. \quad \text{(XIV)}$$

The idealized structure can be interpreted as an OD structure (SEDLACEK, unpublished). Therefore the great number of K_ν-functions may be reduced using the theory of MDO structures.

References

DANIELS, L. K.: Phil. Mag. **14**, 487 (1966)
DORNBERGER-SCHIFF, K., FARKAS-JAHNKE, M.: Acta Cryst. **A26**, 24 (1970)
FARKAS-JAHNKE, M.: Acta Cryst. **A21**, 173 (1966)
FARKAS-JAHNKE, M., DORNBERGER-SCHIFF, K.: Acta Cryst. **A26**, 35 (1970)
FARKAS-JAHNKE, M., GACS, P.: Kristall und Technik **14**, 1475 (1979)
GOMES DE MESQUITA, A. H.: Acta Cryst. **18**, 128 (1965)
JAGNER, S. et al.: Acta Chem. Scand. **A28**, 623 (1974)
JAGNER, S., JANSEN, B.: Acta Cryst. **B33**, 2757 (1977)
KAKINOKI, J. et al.: Acta Cryst. **A25**, 385 (1969)
KAKINOKI, J. et al.: Acta Cryst. **A28**, 292 (1972)
KIHARA, K. et al.: Z. Krist. **147**, 159 (1978)
KRETSCHMAR, R. G.: unpublished
KUBAN, R.-J., SEDLACEK, P.: Kristall und Technik **14**, 1063 (1979)
MEYER, H.-J.: Z. Krist. **128**, 183 (1969)
MÜLLER, U.: Acta Cryst. **A35**, 957 (1979)
SEDLACEK, P.: unpublished (1978)
SEDLACEK, P., DORNBERGER-SCHIFF, K.: Acta Cryst. **18**, 407 (1965)
SEDLACEK, P., DORNBERGER-SCHIFF, K.: Acta Cryst. **B27**, 1532 (1971)
SZULZEWSKY, K.: Dr. Thesis, Akademie der Wissenschaften der DDR, Berlin 1975
TOKONAMI, M.: Miner. J. **4**, 401 (1966)
WALTHER, G.: Collected Abstracts "Symposium on special topics of X-ray crystal structure analysis", Hohengrün 1980

(Received 1) October 23, 1986;
2) December 31, 1986)

Authors' address:

Dr. P. SEDLACEK, Dr. R.-J. KUBAN, Dr. K.-O. BACKHAUS
Zentralinstitut für physikalische Chemie
der Akademie der Wissenschaften der DDR
Rudower Chausee 5
Berlin
DDR-1199

Cryst. Res. Technol.	22	1987	7	929–941

M. T. SEBASTIAN, P. KRISHNA

Physics Departments, University of Cochin, Cochin, India, and
Banaras Hindu University, Varanasi, India

X-ray Investigation of the Mechanism of Phase Transformation in Single Crystals of ZnS, $Zn_xCd_{1-x}S$ and $Zn_xMn_{1-x}S$ (I)

Calculation of Diffraction Effects by a Three Parameter Model

Hexagonal close-packed (2H) single crystals of ZnS, $Zn_xCd_{1-x}S$ and $Zn_xMn_{1-x}S$ are known to undergo solid state transformation to the cubic close-packed (3C) and 6H structures on annealing at elevated temperatures. The transformations occur by the non-random nucleation of stacking faults on individual close-packed layers parallel to (0001). The nature of the faults involved and their probability distribution during transformation determine the diffraction effects produced along the 10.L reciprocal lattice row by a crystal quenched in an intermediate state of transformation. We have investigated the mechanism of the transformation by comparing the diffraction effects recorded from such crystals on a single crystal diffractometer, with those calculated for an assumed model of the transformation.

It is known that in these materials the faults involved in the transformation are deformation faults. To explain the observed diffraction effects we develop a three parameter theoretical model employing a fault probability α for the random insertion of a deformation fault in the 2H structure, a fault probability β for the deformation faults to occur at three layer separations and a fault probability γ of their occurrence at 2-layer separations. The probability α corresponds to the development of a fresh nucleus, the probability β to the growth of the 6H nucleus and the probability γ to the growth of a 3C nucleus. This paper develops the necessary theory of X-ray scattering for such a model of the transformation and predicts the diffraction effects for different values of α, β, and γ. The next paper compares these results with experimental observations.

1. Introduction

Single crystals of ZnS grown from the vapour at temperatures above 1100 °C are known to display a large variety of crystallographic modifications. The Wurtzite (2H) with ABAB ... packing and the Sphalerite (3C) with ABCABC ... packing are the two simplest forms, related by a reversible phase transformation that occurs around 1020 °C, the Wurtzite form being the high temperature modification. Several workers (MÜLLER; ROTH; ADDAMIANO, AVEN; SECCO E'ARAGONA; SEBASTIAN et al. 1982; SEBASTIAN, KRISHNA 1984a) have investigated the 2H–3C transformation in ZnS. One-dimensionally disordered structures representing intermediate states in a 2H–3C transformation have been often reported (RAI, KRISHNA; FLEET). Recently SEBASTIAN and KRISHNA (1983a, 1984b) have observed a 2H–6H transformation to occur in single crystals of $Zn_xCd_{1-x}S$ (for $x \geqq 0.94$) and $Zn_xMn_{1-x}S$ (for $x \geqq 0.9$).

There has been considerable speculation about the mechanism of these transformations. DANIELS and MARDIX et al. proposed a periodic slip mechanism akin to the pole mechanism suggested earlier for the 2H–3C transformation in alloys (SEEGER). This mechanism envisages the expansion of a deformation fault around an axial screw dislocation. The crystals undergoing transformation by such a mechanism show easily measurable kinks in the external shape of the crystals (MARDIX, STEINBERGER). The kink angle for the 2H–3C transformation is 19° 28′. When transformation occurs by such a mechanism the spots corresponding to the new structure would appear on X-ray diffraction photographs at a very early stage of the transformation, and there would be almost no diffuse streak along the reciprocal lattice rows parallel to c^*. Such a mechanism can operate only in crystals which contain an axial screw dislocation and that too, if the mechanism introducing stacking faults does not introduce many other faults which would hinder this mechanism. Not all crystals contain such a dislocation, but all of them undergo the 2H–3C phase transformation on annealing (SEBASTIAN et al. 1982; SEBASTIAN, KRISHNA 1984a). In the absence of a screw dislocation the structural transformations can take place by the statistical insertion of stacking faults which nucleate preferentially at such layer spacings as would lower the energy of the structure. The stacking faults expand rapidly over the entire layer as they have a negative energy. Thus the 2H structure can transform to 3C by the preferential occurrence of deformation faults at two layer separations as depicted below.

```
initial structure (2H):    A B | A B A B A B ...
                                 C A | C A C A ...
                                       B C | B C ...
                                             A B ...

resulting structure (3C):  A B C A B C A B ...
```

It was observed by SECCO D'ARAGONA et al. in the electron microscope that the 2H–3C transformation in ZnS occurs by such a mechanism.

Similarly the 2H–6H transformation can take place by the preferential insertion of deformation faults at 3-layer separations as shown below:

```
initial structure (2H):    A B | A B A B A B A B A B ...
                                 C A C | A C A C A C A ...
                                         B A B | A B A B ...
                                                 C A C | A ...
                                                         B ...

resulting structure (6H):  A B C A C B A B C A C B ...
```

In all such cases since the nucleation of faults is a statistical process, the end product would invariably be a heavily disordered structure containing a high concentration of stacking faults.

Six slip vectors $\pm S_i$ ($i = 1, 2, 3$) describe stacking faults in terms of slip vectors of the type $a/3$ [1$\bar{1}$00]. Of the six slip vectors $\pm S_i$, only three are possible on a particular layer if the rules of the close-packing are not to be violated. A slip vector $+S_i$ produces a cyclic shift of the layer A → B → C → A. While the slip vectors $-S_i$ lead to an antihelic shift of layers A → C → B → A.

If the transformation occurs by the expansion of deformation faults on helical basal planes (created by an axial screw dislocation) then the slip vector is exactly the same every time a partial dislocation bounding the fault goes once around the screw. Therefore, the resulting 3C structure will be untwinned and will occur in kinked regions within the 2H matrix. This has been observed by MARDIX and STEINBERGER who have also measured the kink angle and shown it to be equal to the theoretically calculated value of 19° 28′. If a number of different deformation faults nucleate in different parts of such a 2H crystal and simultaneously expand over the helical planes they can give rise to cubic regions in twin orientations, finally producing a twinned cubic crystal. However, since the expansion of faults in crystals containing a screw dislocation involves only the glide of partial dislocations over the helical basal plane, it would be a rapid process, and one would expect to obtain macroscopic kinked regions of untwinned 3C structure. This means that the concentration of twin faults in the final 3C structure would be small and untwinned domains would be large. In general, this will produce twin reflections of unequal intensity on X-ray diffraction photographs and only very weak diffuse scattering along the reciprocal lattice rows parallel to c^*. In the crystals which we studied (SEBASTIAN et al. 1982; SEBASTIAN, KRISHNA 1984a) this was not the case. The twin reflections are broad and of comparable intensity indicating a large density of twin faults. The crystals under examination under an optical microscope did not reveal any macroscopic kinks.

When the 2H crystals do not contain an axial screw dislocation they cannot transform by the periodic slip mechanism. The basal planes are not helical but flat and faults have to nucleate on individual layers and spread over the entire basal plane in order to effect the transformation. Since each deformation fault nucleates afresh, the partials bounding them can have Burgers vectors which have equal probabilities of lying along any of the three symmetry related directions S_i where

$$S_1 = \frac{a}{3}\,[1\bar{1}00]\,, \qquad S_2 = \frac{a}{3}\,[01\bar{1}0]\,, \quad \text{and} \quad S_3 = \frac{a}{3}\,[\bar{1}010]\,.$$

This would produce no macroscopic kinks. The 2H–3C transformation can thus occur by the non-random nucleation of deformation faults predominantly on alternate layers. Initially the deformation faults would nucleate in different regions of the 2H structure in a random manner. Therefore in the initial stages of the transformation the 2H crystal would contain a random distribution of deformation faults. These then act as nuclei for the cubic phase which grow during subsequent transformation into thick cubic regions by the preferential insertion of deformation faults at two layer separations. The resulting 3C structure finally contains a random distribution of twin faults since the 3C nuclei have equal probability of being in any one of the two twin orientations namely ABCABC ... or ACBACB ... These predictions have been verified (SEBASTIAN, KRISHNA 1984 a, c) and the 2H crystal is found to contain a random distribution of deformation faults in the initial stages of the transformation and a random distribution of twin faults in the resulting 3C structure after transformation. At an intermediate stage of the transformation the theories based on random faulting break down, and the theory of X-ray diffraction from non-randomly faulted crystals has to be applied.

2. Principle of the method of investigation

Since the faults occur non-randomly during transformation, the intensity distribution along the reciprocal lattice rows ($H - K \neq 3n$) perpendicular to the layers cannot be obtained from the conventional theories of X-ray scattering by one -dimensionally disordered close-packed structures which assume a random distribution of faults. The theories for random faulting would only be applicable either in the very early stages of the transformation or when the transformation is nearly completed (SEBASTIAN, KRISHNA 1984c). This is because the initial distribution of the nuclei of the second phase would be random and so will be the distribution of faults in the end product when these nuclei have grown through non-random faulting to meet at faulted interfaces. In order to compute the diffraction effects at an intermediate stage of transformation a more complex fault probability distribution dependent on the mechanism of solid state transformation, has to be employed. Therefore, although X-ray diffraction techniques describe only the average structure, it is possible to obtain useful information about the mechanism of solid state transformations in crystals by arresting the transformation in an intermediate state and examining the diffraction effects produced by it. If we have a model of the disordered intermediate state we can calculate the diffuse intensities and diffraction effects expected from it and compare these with those actually observed, to determine if the proposed model is correct. As we shall see this method can be applied quite successfully to study the mechanism of solid state transformations which involve only one dimensional disorder. In the case of two or three dimensionally disordered systems the theoretical computation would become much more complex.

3. The three parameter model

It is known from an analysis of the broadening of slightly faulted 2H crystals that the faults which occur in ZnS are predominantly deformation faults (ROTH; SEBASTIAN et al. 1982). If these occur preferentially at two layer separations they give rise to a 2H–3C transformation and if they occur preferentially at three layer separations they give rise to the 2H–6H transformation. However, a close scrutiny of X-ray diffractograms shows that there is always an enhancement of diffracted intensity in positions of 6H reflections during a 2H–3C transformation and conversely, there is an enhancement of intensity in positions of 3C reflections during the 2H–6H transformation. In this paper we therefore employ a general three parameter model with the following assumptions:

(i) The crystal is infinite in size and free of distortions.

(ii) The scattering power of all the layers is the same. In the case of ZnS each 'layer' of structure is a double layer of Zn and S atoms displaced relative to each other along [001].

(iii) There is no change in the layer spacings at the faults.

(iv) The faults extend right across the crystal.

(v) The probability for faults to nucleate at random at large separations is α, which is different from the probability β for the faults to occur at 3-layer separations and the probability γ for them to occur at 2-layer separations. The probability for faults to occur on successive layers is negligible and may be equated to zero since its occurrence on every layer do not give rise to a 2H or 6H structure but to a 2H itself.

4. The theory of x-ray diffraction

Employing the notations used earlier (PANDEY et al.; SEBASTIAN, KRISHNA 1984d) we distinguish eight types of layers in the crystal undergoing the transformation by the deformation mechanism. In the perfect 2H structure there are two types of layers with subscripts 0 and 1 according as the stacking offset vector is $+S_i$ or $-S_i$. The perfect 2H structure can thus be written as

$$A_0B_1A_0B_1A_0B_1A_0B_1 \ldots$$

We shall assign the subscripts 0^1 and 1^1 to the first layer of the slipped stack according as it follows a 0 type or 1 type layer. The layer succeeding the 0^1 or 1^1 type layer has been assigned subscripts 0^2 or 1^2 since the faults occur with a probability γ causing the growth of 3C nuclei from this type of layers. The layers succeeding the 0^2 or 1^2 are designated as 0^3 or 1^3 since the 6H nucleus grows with a probability β from these layers. We can now compute the probability $P_{(m,j)}$ of arriving at the mth layer with subscript j ($j = 0, 1, 0^1, 1^1, 0^2, 1^2, 0^3$ or 1^3) with the help of the following probability trees which consider the transitions from $(m-1)^{\text{th}}$ layer to the mth layer with each type of layer at the origin.

The probability tree for the crystals undergoing solid state transformation by the above three-parameter model may be written as follows:

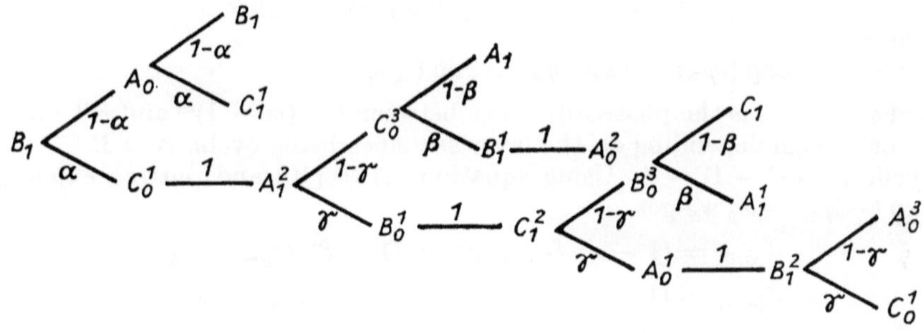

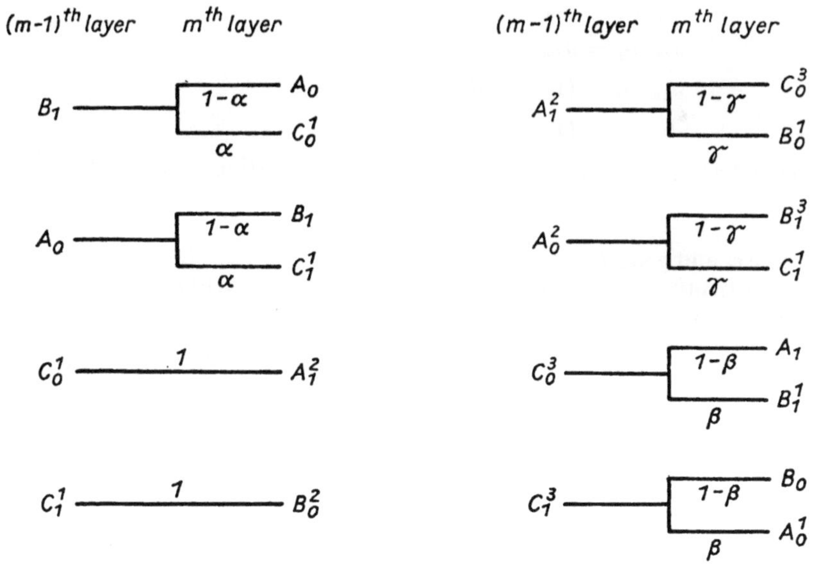

As shown in the probability trees, the layer type following B_1 (or A_0) layers in the absence of a deformation fault is A_0 (or B_1) and occurs with the probability $(1-\alpha)$. Therefore in the presence of a deformation fault the layer following B_1 (or A_0) is C_0^1 (or C_1^1).

The probability $P_{(m,j)}$ may now be computed as follows:

$$P_{(m,0)} = (1-\alpha)\,P_{(m-1,1)} + (1-\beta)\,P_{(m-1,1^3)}\,, \tag{1}$$

$$P_{(m,1)} = (1-\alpha)\,P_{(m-1,0)} + (1-\beta)\,P_{(m-1,0^3)}\,, \tag{2}$$

$$P_{(m,0^1)} = \alpha P_{(m-1,1)} + \beta P_{(m-1,1^3)} + \gamma P_{(m-1,1^2)}\,, \tag{3}$$

$$P_{(m,1^1)} = \alpha P_{(m-1,0)} + \beta P_{(m-1,0^3)} + \gamma P_{(m-1,0^2)}\,, \tag{4}$$

$$P_{(m,0^2)} = P_{(m-1,1^1)}\,, \tag{5}$$

$$P_{(m,1^2)} = P_{(m-1,0^1)}\,, \tag{6}$$

$$P_{(m,0^3)} = (1-\gamma)\, P_{(m-1,1^2)}\,, \tag{7}$$

$$P_{(m,1^3)} = (1-\gamma)\, P_{(m-1,0^2)}\,. \tag{8}$$

The function $J_{(m,j)}$ can be written (PRASAD, LELE) as

$$J_{(m,j)} = \langle \exp (i\varphi_m)_j \rangle = \sum P_{(m,j)} \exp (i\varphi_m)_j\,, \tag{9}$$

where

$$\exp (i\varphi_m) = \exp (i\varphi_{m-1}) \exp (\pm i\varphi_0) \tag{10}$$

and $\varphi_0 = 2\pi/3$ is the phase difference between the $(m-1)^{\text{th}}$ and m^{th} layers and takes + or − sign depending on the layer sequence being cyclic A → B → C → A or anticyclic A → C → B → A. Using equation (1) to (10) and putting $\exp (i\varphi_0) = \omega_1$ and $\exp (-i\varphi_0) = \omega_2$ we get

$$J_{(m,0)} = (1-\alpha)\, J_{(m-1,1)}\omega_2 + (1-\beta)\, J_{(m-1,1^3)}\omega_2 \tag{11}$$

$$J_{(m,1)} = (1-\alpha)\, J_{(m-1,0)}\omega_1 + (1-\beta) J_{(m-1,0^3)}\omega_1\,, \tag{12}$$

$$J_{(m,0^1)} = \alpha J_{(m-1,1)}\omega_1 + \beta J_{(m-1,1^3)}\omega_1 \gamma J_{(m-1,1^2)}\omega_1\,, \tag{13}$$

$$J_{(m,1^1)} = \alpha J_{(m-1,0)}\omega_2 + \beta J_{(m-1,0^3)}\omega_2 + \gamma J_{(m-1,0^2)}\omega_2\,, \tag{14}$$

$$J_{(m,0^2)} = J_{(m-1,1^1)}\omega_2\,, \tag{15}$$

$$J_{(m,1^2)} = J_{(m-1,0^1)}\omega_1\,, \tag{16}$$

$$J_{(m,0^3)} = (1-\gamma)\, J_{(m-1,1^2)}\omega_2\,, \tag{17}$$

$$J_{(m,1^3)} = (1-\gamma)\, J_{(m-1,0^2)}\omega_1\,. \tag{18}$$

Let the solution of the system of difference equations (11) to (18) be of the form

$$J_{(m,j)} = c_j \varrho^m \qquad m \geqq 0 \tag{19}$$

where c_j and ϱ are functions of α, β, and γ.

Substituting in equations (11) to (18) and eliminating the various c's we finally get the so called characteristic equation

$$\begin{aligned} &\varrho^8 + [\gamma - (1-\alpha)^2]\,\varrho^6 + [\alpha(1-\beta)(1-\gamma) + \gamma^2 - \gamma(1-\alpha)^2]\,\varrho^4 + \\ &+ [\beta^2(2\gamma-1) + 2\alpha\gamma(1-\beta) - \gamma^2 - \gamma^2(\alpha-\beta)^2]\,\varrho^2 + \\ &+ (1-\gamma)^2(\alpha-\beta)^2 = 0\,. \end{aligned} \tag{20}$$

HOLLOWAY obtained an analytic solution for the diffracted intensity in terms of coefficients and boundary conditions of the characteristic equation. His intensity expression

$$I = f^2 c \left(\frac{1}{2} + \frac{\sum\limits_{j=1}^{n-1} \sum\limits_{k=0}^{j-1} a_{n-k} T_{j-k} \exp (n-j)^{i\pi L} - a_0}{\sum\limits_{j=0}^{n} a_j \exp (ji\pi L)} \right) + \begin{array}{l}\text{complex} \\ \text{conjugate,}\end{array} \tag{21}$$

where a_n is the coefficient of ϱ^n in the characteristic equation. T's are the boundary conditions, c is a scale factor and

$$f^2 = f_{\text{Zn}}^2 + f_{\text{S}}^2 + 2 f_{\text{Zn}} f_{\text{S}} \cos \frac{3\pi L}{4}\,.$$

In the present case

$$n = 8\,, \quad a_8 = 1\,, \quad a_6 = \gamma - (1-\alpha)^2\,,$$
$$a_4 = \alpha(1-\beta)(1-\gamma) + \gamma^2 - \gamma(1-\alpha)^2\,,$$
$$a_2 = \beta^2(2\gamma - 1) + 2\alpha\gamma(1-\beta) - \gamma^2[1 + (\alpha-\beta)^2]\,,$$
$$a_0 = (1-\gamma)^2(\alpha-\beta)^2\,.$$

The boundary conditions (T's) can be found out by the following method. First we find the probability W_j of finding a layer with subscript j on passing through an arbitrary region of the crystal. From the probability tree we get

$$W_0 = (1-\alpha)\,W_1 + (1-\beta)\,W_1^3\,,$$
$$W_1 = (1-\alpha)\,W_0 + (1-\beta)\,W_0^3\,,$$
$$W_0^1 = \alpha W_1 + \beta W_1^3 + \gamma W_1^2\,,$$
$$W_1^1 = \alpha W_0 + \beta W_0^3 + \gamma W_0^2\,,$$
$$W_1^2 = W_0^1\,,$$
$$W_0^2 = W_1^1\,,$$
$$W_0^3 = (1-\gamma)\,W_1^2\,,$$
$$W_1^3 = (1-\gamma)\,W_0^2\,,$$
$$W_0 + W_1 + W_0^1 + W_1^1 + W_1^2 + W_0^2 + W_0^3 + W_1^3 = 1\,.$$

Solving the above equations we get

$$W_0 = W_1 = \frac{1-\beta-\gamma+\beta\gamma}{2(1+3\alpha-\beta-\gamma+\beta\gamma-\alpha\gamma)}\,,$$

$$W_0^1 = W_1^1 = W_0^2 = W_1^2 = \frac{\alpha}{2(1+3\alpha-\beta-\gamma+\beta\gamma-\alpha\gamma)}\,,$$

$$W_0^3 = W_1^3 = \frac{\alpha(1-\gamma)}{2(1+3\alpha-\beta-\gamma+\beta\gamma-\alpha\gamma)}\,.$$

The boundary conditions can now be obtained by considering all possible sequences starting with layers of each type 0, 1, 0^1, 1^1, 0^2, 1^2, 0^3 and 1^3 at the origin and writing T_m in each case i.e.

$$T_m = \sum_j W_j < \exp(i\varphi_m)_j\,.$$

Thus

$$T_1 = -\tfrac{1}{2}\,,$$

$$T_2 = \frac{1-\beta-\gamma+\beta\gamma-\alpha\gamma}{1+3\alpha-\beta-\gamma+\beta\gamma-\alpha\gamma}\,,$$

$$T_3 = \frac{-1+\beta+\gamma-\beta\gamma+4\alpha\gamma}{2(1+3\alpha-\beta-\gamma+\beta\gamma-\alpha\gamma)}\,,$$

$$T_4 = \frac{1-3\alpha-\beta-\gamma+3\alpha\beta+2\alpha\gamma+\beta\gamma-3\alpha\beta\gamma}{1+3\alpha-\beta-\gamma+\beta\gamma-\alpha\gamma}\,,$$

$$T_5 = (-1+3\alpha+\beta+\gamma-\beta\gamma+\alpha\gamma+3\alpha^2-6\alpha\beta-6\alpha\gamma^2+6\alpha\beta\gamma - \\ - 3\alpha^2\beta - 3\alpha^2\gamma + 3\alpha^2\beta\gamma)/2(1+3\alpha-\beta-\gamma+\beta\gamma-\alpha\gamma)\,,$$

$$T_6 = (2 - 12\alpha - 2\beta - 2\gamma + 18\alpha\beta + 10\alpha\gamma + 2\beta\gamma + 12\alpha^2 - 6\alpha^3 - - 24\alpha\beta\gamma - 12\alpha^2\beta - 12\alpha^2\gamma + 6\alpha\gamma^2 + 6\alpha^3\gamma + 12\alpha^2\beta\gamma + 6\alpha^3\beta + + 6\alpha\beta^2\gamma + 6\alpha\beta\gamma^2 - 6\alpha\beta^2\gamma - 6\alpha^3\beta\gamma)/2(1 + 3\alpha - \beta - \gamma + \beta\gamma - \alpha\gamma)$$

$$T_7 = (-1 + 6\alpha + \beta + \gamma - 3\alpha^2 - \beta\gamma - 12\alpha\beta - 2\alpha\gamma + 18\alpha\beta\gamma + 3\alpha\beta^2 - - 9\alpha^2\gamma + 3\alpha^2\beta - 9\alpha\gamma^2 + 6\alpha\beta\gamma^2 + 3\alpha\gamma^3 + 3\alpha^4 + 12\alpha^2\gamma^2 + 9\alpha^2\beta\gamma - - 6\alpha\beta^2\gamma - 3\alpha^4\gamma - 12\alpha^2\beta\gamma^2 - 3\alpha^4\beta - 3\alpha^3\beta^2\gamma + 3\alpha\beta^2\gamma^2 + 3\alpha^2\beta^2\gamma)/ 2(1 + 3\alpha - \beta - \gamma + \beta\gamma - \alpha\gamma) .$$

Substituting the coefficients (a's) and the seven boundary conditions (T's) in equation (21) and carrying out the summations we get the intensity expression as

$$I = f^2c\left[\left(\frac{1}{2} + \frac{P}{Q}\right) + \text{complex conjugate}\right], \tag{24}$$

$$P = T_1e^{7i\pi L} + T_2e^{6i\pi L} + (T_3 + a_6T_1)\,e^{5i\pi L} + (T_4 + a_6T_2)\,e^{4i\pi L} + + (T_5 + a_6T_3 + a_4T_1)\,e^{3i\pi L} + (T_6 + a_6T_4 + a_4T_2)\,e^{2i\pi L} + (T_7 + a_6T_5 + a_4T_3 + a_2T_1)\,e^{i\pi L} - a_0\,,$$

$$Q = a_0 + a_2e^{2i\pi L} + a_4e^{4i\pi L} + a_6e^{6i\pi L} + e^{8i\pi L} .$$

On simplifying equation (24) we get the final intensity expression for a 2H crystal undergoing solid state transformation as

$$I = f^2c\left(1 + \frac{x}{y}\right) \tag{25}$$

where

$$x = B_1R_1 + a_0B_1R_7 + a_2B_1R_5 + a_4B_1R_3 + a_6B_1R_1 + B_2R_2 + a_0B_2R_6 + + a_2B_2R_4 + a_4B_2R_2 + a_6B_2 + B_3R_3 + a_0B_3R_5 + a_2B_3R_3 + + a_4B_3R_1 + a_6B_3R_1 + B_4R_4 + a_0B_4R_4 + a_2B_4R_2 + a_4B_4 + + a_6B_4R_2 + B_5R_5 + a_0B_5R_3 + a_2B_5R_1 + a_4B_5R_1 + a_6B_5R_3 + + B_6R_6 + a_0B_6R_2 + a_2B_6 + a_4B_6R_2 + a_6B_6R_4 + B_7R_7 + + a_0B_7R_1 + a_2B_7R_1 + a_4B_7R_3 + a_6B_7R_5 + B_8R_8 + a_0B_8 + + a_2B_8R_2 + a_4B_8R_4 + a_6B_8R_6 .$$

$$y = G_1 + G_2R_2 + G_3R_4 + G_4R_6 + G_5R_8 + G_6R_2 + G_7R_4 + G_8R_6 + + G_9R_2 + G_{10}R_4 + G_{11}R_2\,,$$

$$B_1 = 2T_1\,,$$
$$B_2 = 2T_2\,,$$
$$B_3 = 2(T_3 + a_6T_1)\,,$$
$$B_4 = 2(T_4 + a_6T_2)\,,$$
$$B_5 = 2(T_5 + a_6T_3 + a_4T_1)\,,$$
$$B_6 = 2(T_6 + a_6T_4 + a_4T_2)\,,$$
$$B_7 = 2(T_7 + a_6T_5 + a_4T_3 + a_2T_1) .$$
$$B_8 = -2a_0\,,$$

$$\begin{aligned}
R_1 &= \operatorname{Cos} \pi L\,, & G_1 &= 1 + a_0^2 + a_2^2 + a_4^2 + a_6^2\,,\\
R_2 &= \operatorname{Cos} 2\pi L\,, & G_2 &= 2a_0a_2\,,\\
R_3 &= \operatorname{Cos} 3\pi L\,, & G_3 &= 2a_0a_4\,,\\
R_4 &= \operatorname{Cos} 4\pi L\,, & G_4 &= 2a_0a_6\,,\\
R_5 &= \operatorname{Cos} 5\pi L\,, & G_5 &= 2a_0\,,\\
R_6 &= \operatorname{Cos} 6\pi L\,, & G_6 &= 2a_2a_4\,,\\
R_7 &= \operatorname{Cos} 7\pi L\,, & G_7 &= 2a_2a_6\,,\\
R_8 &= \operatorname{Cos} 8\pi L\,, & G_8 &= 2a_2\,,\\
 & & G_9 &= 2a_4a_6\,,\\
 & & G_{10} &= 2a_4\,,\\
 & & G_{11} &= 2a_6\,.
\end{aligned}$$

5. Results and discussion

The variation of the diffracted intensity with $L(h_3)$ in reciprocal space for different values of α, β, and γ as calculated from equation (25) in steps of $\Delta L = 0.01$ are depicted in Figure 1 for the 2H–6H transformation and in Figure 2 for the 2H–3C

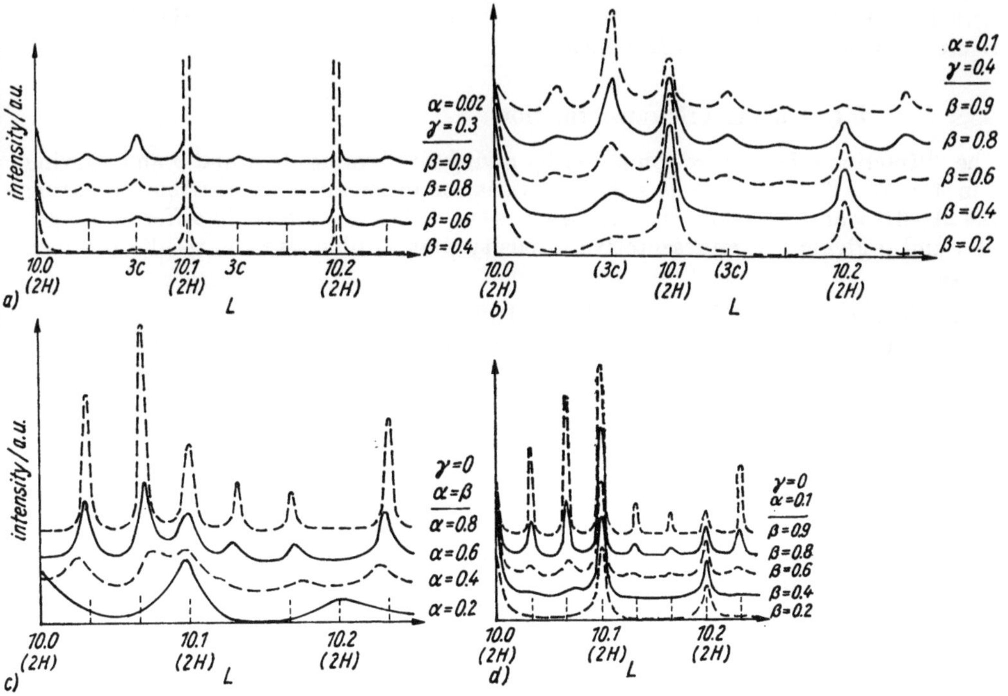

Fig. 1. Calculated variation of the diffracted intensity along the 10.L reciprocal lattice row for a 2H crystal undergoing transformation by the three parameter model for different values of α, β, and γ. The calculated profiles are shifted vertically for clarity: (a) $\alpha = 0.02$, $\gamma = 0.3$, $\beta = 0.4, 0.6, 0.8, 0.9, 0.95$; (b) $\alpha = 0.1$, $\gamma = 0.4$, $\beta = 0.2, 0.4, 0.6, 0.8, 0.9$; (c) $\alpha = \beta = 0.2, 0.4, 0.6, 0.8$, $\gamma = 0$; (d) $\alpha = 0.1$, $\gamma = 0$, $\beta = 0.2, 0.4, 0.6, 0.8, 0.9$ (The vertical dashed lines indicate the 6H positions)

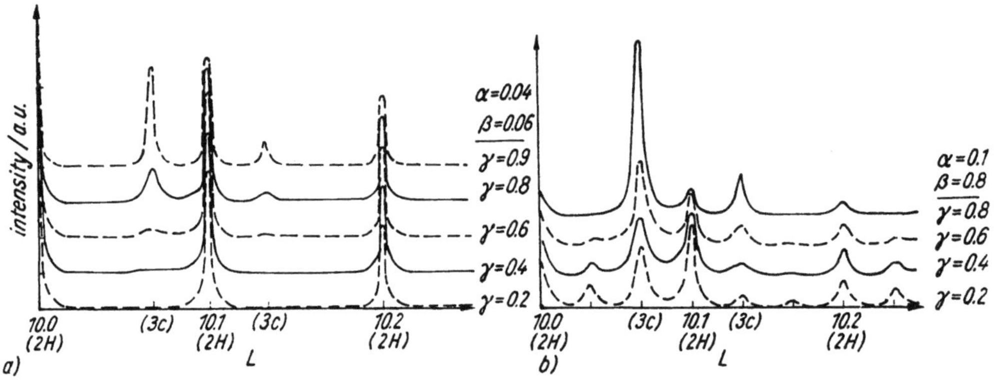

Fig. 2. Calculated variation of the diffracted intensity along the 10.L reciprocal lattice row for a 2H crystal undergoing transformation by the three parameter model for different values of α, β, and γ. The calculated profiles are shifted vertically for clarity: (a) $\alpha = 0.04$, $\beta = 0.06$, $\gamma = 0.2, 0.4, 0.6, 0.8, 0.9$; (b) $\alpha = 0.1$, $\beta = 0.8$, $\gamma = 0.2, 0.4, 0.6, 0.8$

transformation. The reflections along reciprocal lattice rows with $H - K = 0 \bmod 3$ are unaffected by the transformations and remain sharp throughout. In computing the diffracted intensity from equation (25) it is important to note that f^2 is not a constant and varies along the 10.L row and must be taken into account. For reflections with $H - K \neq 0 \bmod 3$ the different diffraction effects are clearly noticeable in the figures and can be described as follows.

5.1. The 2H–6H transformation

The diffraction effects were computed by giving fixed values to α and γ and varying β in steps of 0.2, 0.4, 0.6 etc. to study the variation in the intensity distribution during the 2H–6H transformation. We shall discuss the diffraction effects in different ranges of α and γ giving the representative intensity plot to illustrate the results for each range.

(i) In the range $\alpha = 0.02$ to 0.1 and $\gamma = 0.02$ to 0.4 (e.g. see Fig. 1a).

The 2H reflections remain almost unbroadened. New broadened reflections start appearing at positions corresponding to the 6H reflections as β increases to 0.6. The 6H reflections are almost unshifted and become sharper as β increases to 0.95.

(ii) In the range $\alpha = 0.1$ to 0.2 and $\gamma = 0.1$ to 0.6 (e.g. see Fig. 1b).

The 2H reflections are broadened. Heavily broadened new 6H reflections appear as β increases to a value between 0.4 and 0.6. The 6H reflections with $L = \pm\frac{1}{3}, \pm\frac{5}{3}$, and $\pm\frac{7}{8}$ (which do not coincide with the 3C reflections) show a shift in peak positions (Fig. 1b). The resultant 6H structures shows considerable intensity enhancement near the 3C positions.

The following special cases are of interest:

(a) $\alpha = \beta$ and $\gamma = 0$.

In this case the calculated intensity distribution along the 10.L reciprocal lattice row is shown in Figure 1c for different values of α ($= \beta$). This corresponds to the one parameter model of deformation faulting employed by PANDEY et al. in their investigation of the 2H–6H transformation in SiC. In this model it was assumed that the probability for faults to occur at 1 or 2 layer separations is zero and that the probability for them to occur at any other separation is the same. The co-existence of 2H and 6H peaks during transformation is not predicted by such a model; it is therefore inapplicable in those cases where the 2H and 6H peaks are both present during the transformation.

(b) $\alpha \neq \beta$ but $\gamma = 0$

The calculated intensity distribution along the $10.L$ row is shown in Figure 1d for this case for a small value of α and increasing values of β. This corresponds to the two parameter model for the 2H–6H transformation developed by SEBASTIAN and KRISHNA (1983b, 1984d). This model assumes that the probability of growth of a 6H nucleus is greater than the probability of formation a of fresh nucleus. In this case the diffraction peaks corresponding to the 6H phase develop at an early stage of the transformation and are shifted away from each other. They co-exist with the 2H peaks and rapidly attain their correct positions as the transformation proceeds.

5.2. The 2H–3C transformation

The diffraction effects were computed by giving fixed values to α and β and varying γ in steps of 0.2, 0.4, 0.6 etc. to study the variation in the intensity distribution during the 2H–3C transformation. We discuss below the calculated diffraction effects in different ranges of α, β, and γ giving one representative intensity plot to illustrate the results for each range:

(i) In the range $\alpha = 0.01$ to 0.1, $\beta = 0.02$ to 0.08 (e.g. see Fig. 2a).

The 2H reflections remain sharp and new reflections start appearing at the 3C positions as γ increases to 0.6. The 3C peaks are initially broad and become sharper as γ increases to 0.9. The resultant structure at $\gamma = 0.9$ shows a co-existence of the 2H and 3C peaks (Fig. 2a).

(ii) In the range $\alpha = 0.01$ to 0.2 and $\beta = 0.1$ to 0.8 (e.g. see Fig. 2b).

The 2H reflections remain unbroadened except for large values of α ($\alpha \geqq 0.1$). New reflections appear near the 3C and 6H positions. Those 6H reflections ($L = \pm\frac{1}{3}, \pm\frac{5}{3}$ or $\pm\frac{7}{3}$) which do not coincide with the 3C reflections show peak shifts. As γ increases these reflections weaken and shift towards the nearby 3C reflection (Fig. 2b).

The following special cases are of interest:

(a) $\alpha = \beta \neq \gamma$ (see Fig. 3).

The calculated intensity distribution in this case is shown in Figure 3 and corresponds to the two parameter model for a 2H–3C transformation developed by LELE and PANDEY. When the probability γ for the growth of the 3C nucleii is much larger than the probability α for the creation of a fresh cubic nucleus, the crystal will finally transform into a twinned cubic structure with a large size of individual twin domains. If γ is not much larger than α then the number of twin nucleii will be smaller and the size of individual domains correspondingly larger. From the Figures 3 a, b, and c we make the following observations:

(i) In the range $\alpha = \beta = 0.02$ to 0.06: The 2H reflections remain almost unbroadened. The new 3C reflections start appearing when γ reaches a value between 0.6 and 0.8 (Fig. 3a) and show no peak shifts.

(ii) In the range $\alpha = \beta = 0.08$ to 0.1 (see Fig. 3b): The 2H reflections are broadened and the new 3C reflections with $L = \pm\frac{2}{3}, \pm\frac{4}{3}$ are initially shifted towards the $L = \pm 1$ position (Fig. 3b). The shift decreases as the transformation progresses and the 3C reflections become sharper.

(iii) In the range $\alpha = \beta = 0.2$ to 0.4 (see Fig. 3c): Initially the 2H reflections are heavily broadened ($\gamma = 0.2$) and show large peak shifts. As the transformation progresses the intensity divides into broad 3C peaks which are shifted towards each other

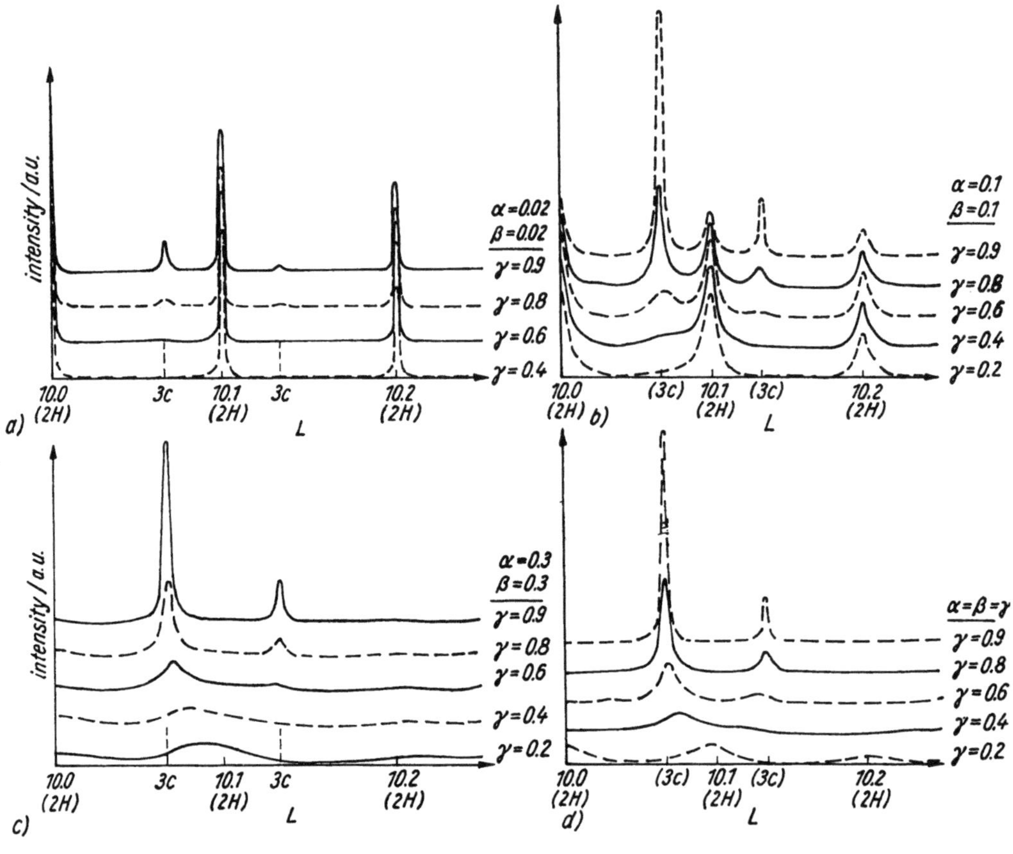

Fig. 3. Calculated variation of the diffracted intensity along the 10.L reciprocal lattice row for a 2H crystal undergoing transformation to the 3C structure by the two parameter model i.e., $\beta = \alpha$, for different values of α with $\gamma = 0.2, 0.4, 0.6, 0.8, 0.9$. The profiles are shifted vertically for clarity for different values of γ: (a) $\alpha = \beta = 0.02$; (b) $\alpha = \beta = 0.1$; (c) $\alpha = \beta = 0.3$; (d) $\alpha = \beta = \gamma$

(e.g. at $\gamma = 0.6$). Finally the 3C peaks migrate to their correct positions $L = \pm \frac{2}{3}$ and $\pm \frac{4}{3}$ as γ increases to 0.9.

(iv) $\alpha = \beta \gg \gamma$: This case is physically meaningless since it does not correspond to a 2H–3C transformation at all: it corresponds to the case of a 2H–6H transformation by the one parameter model with a small fault probability γ at two layer separations. Therefore this was not computed.

(b) $\alpha = \beta = \gamma$: This corresponds to a one parameter model for the 2H–3C transformation which assumes that the probability for faults to occur at one layer separation is zero and is same for all other layer separations. The intensity distribution for such a model is shown in Figure 3d. At no stage of the transformation is there a co-existence of the 2H and 3C reflections. This model is therefore not applicable to those transformations where the 2H and 3C reflections are observed to co-exist during the transformation. One has, however, to be careful to verify that the co-existence of 2H and 3C peaks is not due to untransformed regions in the crystal.

The 2H–3C and 2H–6H transformations have been observed in several materials. In those cases where the transformations occur through a non-random insertion of deformation faults, the three parameter model described above provides a convenient means of computing the diffraction effects for different kinds of fault probability distributions during either transformation. The actual probability distribution obtaining in a particular material undergoing either the 2H–3C or 2H–6H transformations can be determined by comparing the diffraction effects observed experimentally with those given above. This is done in the next paper for the transformations observed in single crystals of ZnS, $Zn_xCd_{1-x}S$, and $Zn_xMn_{1-x}S$. The diffraction effects computed here would also apply to other materials undergoing similar transformations except for the variation in the factor f^2 which depends on the chemical composition and detailed structure of each "layer" of the structure. Before applying the model to a particular material it is necessary to carefully consider whether all the assumptions made in the development of the theory are applicable to the material under investigation.

The authors are grateful to the Department of Science and Technology, New Delhi, for the financial assistance.

References

ADDAMIANO, A., AVEN, M.: J. Appl. Physics **31**, 36 (1960)
DANIELS, B. K.: Phil. Mag. **14**, 487 (1966)
FLEET, M. E.: J. Appl. Cryst. **9**, 190 (1976)
HOLLOWAY, H.: J. Appl. Physics **40**, 4313 (1969)
LELE, S., PANDEY, D.: Materials Sc. Forum **3**, 143 (1985)
MARDIX, S., KALMAN, Z. H., STEINBERGER, I. T.: Acta Cryst. **A24**, 464 (1968)
MARDIX, S., STEINBERGER, I. T.: J. Appl. Physics **41**, 5339 (1970)
MÜLLER, H.: Neues Jahrb. Miner. Abh. **84**, 43 (1952)
PANDEY, D., LELE, S., KRISHNA, P.: Proc. Roy. Soc. **A369**, 435 (1980)
PRASAD, B., LELE, S.: Act. Cryst. **A27**, 54 (1971)
RAI, K. N., KRISHNA, P.: J. Cryst. Growth **3–4**, 741 (1968)
ROTH, W. L.: Faulting in ZnS, General Electric Research Laboratory, Report No. 60-RI-2563m, Schenectady, New York 1960, p. 1–30
SEBASTIAN, M. T., PANDEY, D., KRISHNA, P.: phys. stat. sol. **71**, 633 (1982)
SEBASTIAN, M. T., KRISHNA, P.: Solid State Commun. **48**, 879 (1983a)
SEBASTIAN, M. T., KRISHNA, P.: phys. stat. sol. **79**, 271 (1983b)
SEBASTIAN, M. T., KRISHNA, P.: Pramana **23**, 395 (1984a)
SEBASTIAN, M. T., KRISHNA, P.: J. Cryst. Growth **66**, 586 (1984b)
SEBASTIAN, M. T., KRISHNA, P.: Phil. Mag. **49**, 809 (1984c)
SEBASTIAN, M. T., KRISHNA, P.: Bull. Mat. Sci. **6**, 369 (1984d)
SEEGER, A.: Z. Metalkde. **4**, 247 (1953)
SECCO D'ARAGONA, F. DELAVIGNETTE, P., AMELINCKX, S.: phys. stat. sol. **14**, K115 (1966)

(Received October 11, 1986)

Authors' addresses:

Professor P. KRISHNA
Banaras Hindu University, Department of Physics
Varanasi 221005
India

Dr. M. T. SEBASTIAN*
Physics Department, University of Cochin
Cochin 682022
India

**Present address*:

Lab. de Mineralogie-cristallograpie, Universités Pierre et Marie Curie, 4 Place Jussieu
75252 Paris, Cedex 05, France

Book Review

L. M. FALIKOV, J. L. MORÁN-LÓPEZ (eds)

Magnetic Properties of Low-Dimensional Systems

Springer-Verlag, Berlin, Heidelberg, New York 1986, 189 Seiten, 126 Abbildungen, DM 90.–,
ISBN 3-540-16261-5

Unser Kenntnisstand über niederdimensionale magnetische Systeme ist von grundlegender Bedeutung für das Verständnis magnetischer Phänomene. Gerade in letzter Zeit sind außergewöhnliche Fortschritte auf diesem Gebiet erzielt worden. Die Ursachen dafür liegen vor allem in der Entwicklung neuer, sehr leistungsfähiger Methoden der spinpolarisierten Spektroskopie, deren Meßdaten mit Hilfe hochentwickelter theoretischer Modelle und Verfahren ausgewertet werden können.

Der vorliegende 14. Band der „Proceedings in Physics" des Springerverlages gibt in 23 Artikeln den Themenkreis eines Workshops vom Januar 1986 in Taxco, Mexico, zu diesem Forschungsgebiet wieder. Vorgestellt werden die Ergebnisse von Arbeiten zur Struktur magnetischer Oberflächen, zu experimentellen Untersuchungsmethoden, zur Statistischen Mechanik von Oberflächen und Grenzflächen, zu dünnen Filmen, kleinen Teilchen und Clustern, quasi-eindimensionalen und quasi-zweidimensionalen Systemen.

Die Artikel spiegeln die enge Verflechtung von Theorie und Experiment wider. Die Weiterentwicklung der Spinpolarisierten-Dichtefunktionaltheorie erlaubt heute die zuverlässige Berechnung der elektronischen und magnetischen Struktur des Grundzustandes kleiner Teilchen und komplexer flächenhafter Systeme aus ersten Prinzipien. Andererseits liefert die einfache Spinfluktuationstheorie oft eine zufriedenstellende Beschreibung der Thermodynamik inhomogener ferromagnetischer Systeme.

Eine Vielfalt spektroskopischer Techniken ist in den letzten Jahren weiterentwickelt worden, so vor allem die spinpolarisierten Arten der Elektronenbeugung, der Photoemission und Bremsstrahlung sowie der Elektronen-Einfang-Spektroskopie. Als weitere Meßverfahren werden Mößbauer-Spektroskopie, Elektronenspinresonanz sowie magnetooptische und magnetometrische Methoden vorgestellt.

Der Tagungsband gibt einen nahezu geschlossenen Überblick über den Wissensstand von 1986 auf dem Gebiet niederdimensionaler magnetischer Systeme. Der Inhalt reflektiert eindrucksvoll die Verallgemeinerungsfähigkeit des itinerant-Modells des Magnetismus und gehört daher in die Hand eines jeden Physikers, der in der Grundlagenforschung des Magnetismus arbeitet, sollte aber auch den Praktiker interessieren.

P. MUTH, H. ESCHRIG

Cryst. Res. Technol.	22	1987	7	943–950

A. Waśkowska, A. Pietraszko, Z. Czapla

Institute for Low Temperature and Structure Research,
Polish Academy of Sciences, Wrocław, Poland and
Institute of Physics, University of Wroclaw, Poland

Optical, X-ray, and Calorimetric Studies of Phase Transitions in ND_4BeF_3

Phase transitions in ND_4BeF_3 crystals at $T_1 = 330$ and $T_2 = 245$ K were studied by specific heat (DSC) and X-ray lattice parameters measurements in function of temperature. These ferroelastic transitions are accompanied by anomalous heat absorption, sudden changes in the unit cell parameters and crystal symmetry. Formation of domain structure in particular phases examined under a polarizing microscope is also described. Domain walls of the type (100), (010), and (001) observable at room temperature indicate on the triclinic symmetry of the Phase II. Above T_1 the crystal becomes orthorhombic. First-order phase transition at T_2 causes the symmetry of the crystal in the Phase III to be monoclinic with the **c**-axis unique. The phase transition diagram is proposed and compared with the phase situation in nondeuterated NH_4BeF_3 crystals.

Introduction

Considerable attention has been currently paid to the physical properties of the hydrogen-bonded ferroelectric/ferroelastic crystals. One of them, NH_4BeF_3, with the phase transition at 250 and 334 K was reported as ferroelastic by Makita and Suzuki, and by Czapla et al. The X-ray studies of lattice parameters as a function of temperature and supported by DSC measurements performed by Lukaszewicz et al. (1983) established that there was another phase transition at 347 K. On heating there were found four phases which change at temperatures $T_1 = 347$, $T_2 = 334$ and $T_3 = 252$ K. A peculiar thermal hysteresis was observed when studying the sequence of the phases, and on cooling it was I → II → IV. Besides, the authors noted that within the same batch there were found two kinds of crystals, the first kind with the phase transitions as above and the second one, where the Phase II was missing at all, i.e. these crystals transformed at $T = 334$ K directly from Phase III to Phase I and this transition was fully reversible (I ⇄ III ⇄ IV). No differences in the lattice parameters or in diffraction pattern at room temperature were observed. Calorimetric and dilatometric measurements performed by Yoshida et al. (1984a) as well as optical studies of external stresses (Yoshida et al. 1984b) confirmed the transition at T_2 and T_3. These results reminded a behaviour of the crystal of the second-type mentioned above, in which the intermediate Phase II did not exist.

In order to understand the dynamics of the phase transitions in NH_4BeF_3 it was necessary to know in detail the crystal structure in all its phases. Due to the pseudo-orthorhombic features of the crystal at room-temperature, several different assignments of the space group such as $P2_1/n$ (Milkova, Poray-Koshits), $P2_12_12_1$ (Tedenac, Cot) and lastly — basing on the optical study of the domain structure — P1 (Yoshida et al. 1984b) have been reported. The X-ray crystal structure of the room-temperature

Phase III, described in the space group Pn was published by WAŚKOWSKA (1983). The structure consists of two hydrogen-bonded chains of $(BeF_4)_n$ tetrahedra running along the **b**-axis. High temperature Phases I and II, both orthorhombic, were reported by WAŚKOWSKA (1985) and phase transition mechanism was described in terms of deformation of the Phase I considered as a prototypic phase with the highest symmetry.

Because of differences in behaviour of the two types of NH_4BeF_3 crystals with respect to the phase sequence and uncertainty in a choice of the space groups, further studies seemed to be necessary. As a part of the project we have prepared the ND_4BeF_3 crystals expecting that both substances should be isostructural and if so, ND_4BeF_3 might give additional information on these questions.

The present paper reports the DSC, optical study, and thermal expansion of deuterated ND_4BeF_3 crystals.

Experimental

The ND_4BeF_3 crystals were obtained by four-fold recrystallization of NH_4BeF_3 dissolved in heavy water. They grew as *c*-platelets elongated in the **b**-direction. The specific heat was measured with the Elmer-Perkin's calorimeter (DSC) in a temperature region between 215 to 390 K.

Optical examinations of ND_4BeF_3 under a polarizing microscope were carried out in the temperature range 220–390 K on the *b*- and *c*-plate specimens free from an external stresses. The experiment was considerably hampered by the irregular terrace-shaped surface of the crystal which unabled us to obtain smooth slices of the samples.

Precise lattice parameters were measured by the BOND method using X-ray single crystal diffractometer of special construction (ŁUKASZEWICZ et al. 1978). High and low temperature attachments enabled us to carry out the experiment in the temperature range from 215 to 360 K and with the temperature stabilization ±0.05 K. The Bragg angles of the following 16 reflections with possibly high θ angles ($54° \leqq \theta \leqq 75°$) were measured for three samples:

I. 058, $05\bar{8}$, 0016, 040
II. 5011, $50\bar{1}\bar{1}$, 608, $60\bar{8}$, 600, 0016
III. $6\bar{2}0$, $6\bar{3}0$, $2\bar{8}0$, $1\bar{5}0$, $3\bar{5}0$, $\bar{3}\bar{5}0$, 440

Table 1
The A_i coefficients of the polynomial $P = \sum_{i=1}^{4} A_i T^{(i-1)}$

Phase	lattice parameter	A_1	A_2	A_3	A_4
I	*a*	6.07319	$-1.90712 \cdot 10^{-3}$	$2.71267 \cdot 10^{-6}$	
	b	4.52296	$4.01346 \cdot 10^{-4}$	$-1.93762 \cdot 10^{-7}$	
	c	13.23654	$-3.05454 \cdot 10^{-3}$	$4.96032 \cdot 10^{-6}$	
II	*a*	6.70211	-0.01070	$4.04295 \cdot 10^{-5}$	$-5.05934 \cdot 10^{-8}$
	b	4.20192	$4.29637 \cdot 10^{-3}$	$-1.58122 \cdot 10^{-5}$	$2.04975 \cdot 10^{-8}$
	c	13.71311	$-9.56787 \cdot 10^{-3}$	$3.57712 \cdot 10^{-5}$	$-4.66049 \cdot 10^{-8}$
	α	89.48135	$5.96058 \cdot 10^{-3}$	$-2.24538 \cdot 10^{-5}$	$2.76177 \cdot 10^{-8}$
	β	100.47007	$-0.11136 \cdot 10^{-4}$	$3.90711 \cdot 10^{-4}$	$-4.52240 \cdot 10^{-7}$
	γ	82.90003	0.07733	$-2.69143 \cdot 10^{-4}$	$3.00343 \cdot 10^{-7}$
III	*a*	4.08813	0.02185	$-9.86055 \cdot 10^{-5}$	$1.49012 \cdot 10^{-7}$
	b	4.49039	$1.25814 \cdot 10^{-3}$	$-2.57556 \cdot 10^{-6}$	$0.00011 \cdot 10^{-8}$
	c	9.05087	0.05025	$-2.12972 \cdot 10^{-4}$	$2.98023 \cdot 10^{-7}$
	γ	118.279	-0.30682	$1.15149 \cdot 10^{-3}$	$-1.43051 \cdot 10^{-6}$

The mean relative accuracy of the lattice parameters determination has been estimated to be $\delta a_i/a_i \approx 4 \cdot 10^{-5}$ ($i = 1$, 2 and 3) and of the axial angles: $\pm 0.01°$.

The measured Bragg angles were put into a common temperature scale, which allowed us to calculate lattice parameters as a function of temperature by using the third-rank polynomials of the form $P = \sum_{i=1}^{4} A_i T^{(i-1)}$. The polynomial coefficients A_i in a regions of subsequent phases are given in Table 1.

Results

The results of specific heat measurements are shown in Figure 1. There is a pronounced anomaly at $T_2 = 245$ K and another one, much smaller at $T_1 = 330$ K. The corresponding transition heats are 209 and 46 cal · mol^{-1}. As compared with nondeuterated crystal, a lack of intermediate Phase II with the space group $P2_12_12_1$, was observed and this indicates on a similarity of ND_4BeF_3 to the nondeuterated NH_4BeF_3 crystals of the second type mentioned above.

Figure 2 illustrates the lattice parameters in a function of temperature. Discontinuities in all curves correspond to the phase transitions at T_1 and T_2 confirming the DSC measurements, i.e. there are only three phases in ND_4BeF_3. All results are given on heating-run, since contrary to NH_4BeF_3, a thermal hysteresis here was observed to be about 1–1.2 K for the phase transition at T_2 and 0.4–0.6 K at T_1, respectively, and did not induce a changes in the phase sequence. A common feature of all these curves is that except for an abrupt jump of lattice parameters at the phase transition point, a temperature changes of a given parameter are not large in a region of particular phases. The **a**-parameter is almost temperature independent in the Phase I; a linear thermal expansion along the **c**-axis is negative in the Phases II and III. The axial angle α equals to $90.000 \pm 0.001°$ in all phases, the angle β differs from the right angle only in the Phase II, whereas $\gamma = 90.10 \pm 0.02°$ and $91.34 \pm 0.04°$ in the Phases II and III, respectively. This results indicate that ND_4BeF_3 crystal is orthorhombic in the Phase I, triclinic at room temperature and most likely monoclinic in the Phase III.

In Figure 3 there is shown a plot of the unit cell volume V (Å^3) against the temperature. The anomalous decrease of the lattice cell volume V with increasing temperature at the phase transition points may be understood in connection with molecular model of the phase transition mechanism (WAŚKOWSKA 1983, 1985). X-ray crystal structure suggested that transition to the low temperature Phases II and III is realized by

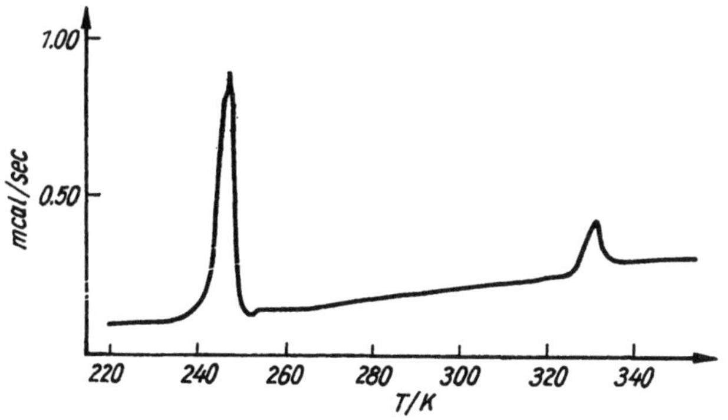

Fig. 1. Differential scanning calorimetry signals for ND_4BeF_3

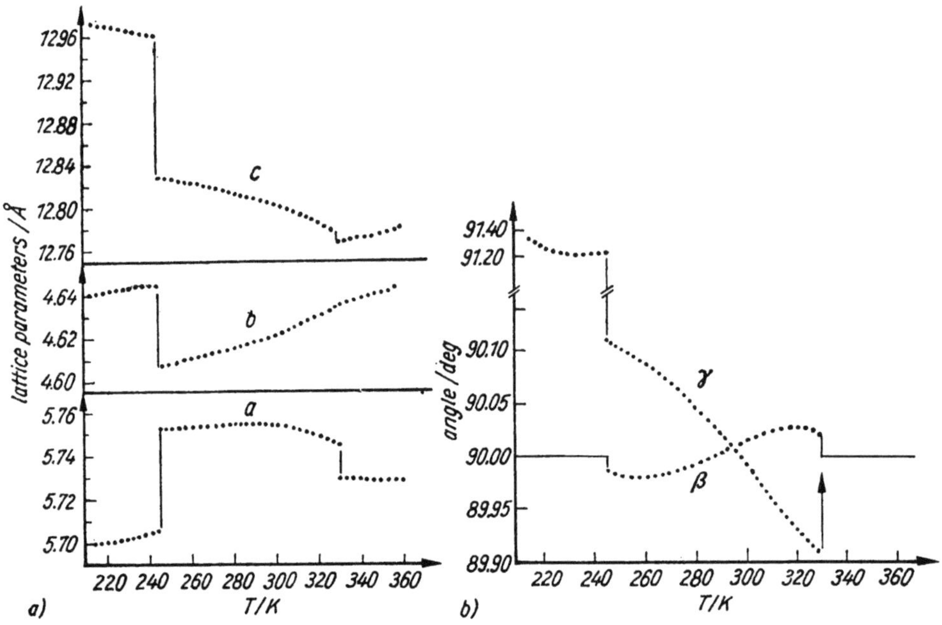

Fig. 2. Linear and axial parameters in function of temperature

deviations of structural elements from the high symmetry positions (prototype phase), i.e. there take place conformational changes in $(BeF_4)_n$ chains causing an increase of mutual distances between the chains in the **c**-direction and elongation of N–F distances when NH_4 ions deviate from their mirror plane symmetry positions.

In order to complete the thermal deformation experiment we have studied a domain formation versus temperature under a polarizing microscope. In Figure 4a is shown the **c**-plate specimen with one central large domain and two smaller ones of the type (100) at room temperature. When the sample was cooled down below T_2 there appeared rapidly new, very narrow domains, typical for ferroelastic twining. In Figure 4b is shown an enlarged fragment of Figure 4a with this type (010) domains. After heating back to room temperature the narrow twin structure disappeared (Fig. 4c). Further heating up above T_1 caused the domains with the (100) walls to disappear also. The sample cooled down back to room temperature, except of the old (100) domains, exhibits an additional twined structure with the wider (010) and (100) domain walls (Fig. 4c).

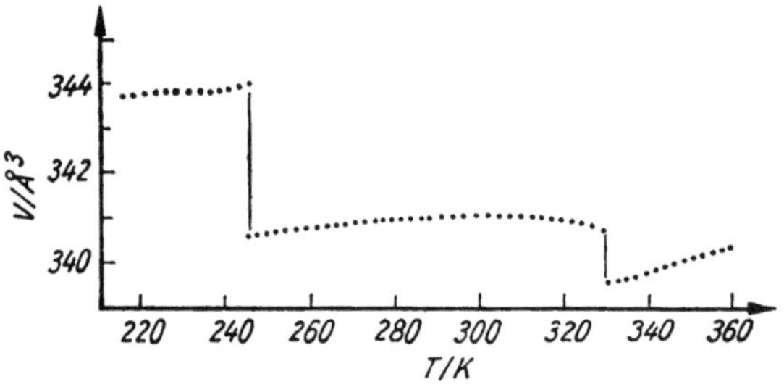

Fig. 3. Temperature dependence of the unit cell volume

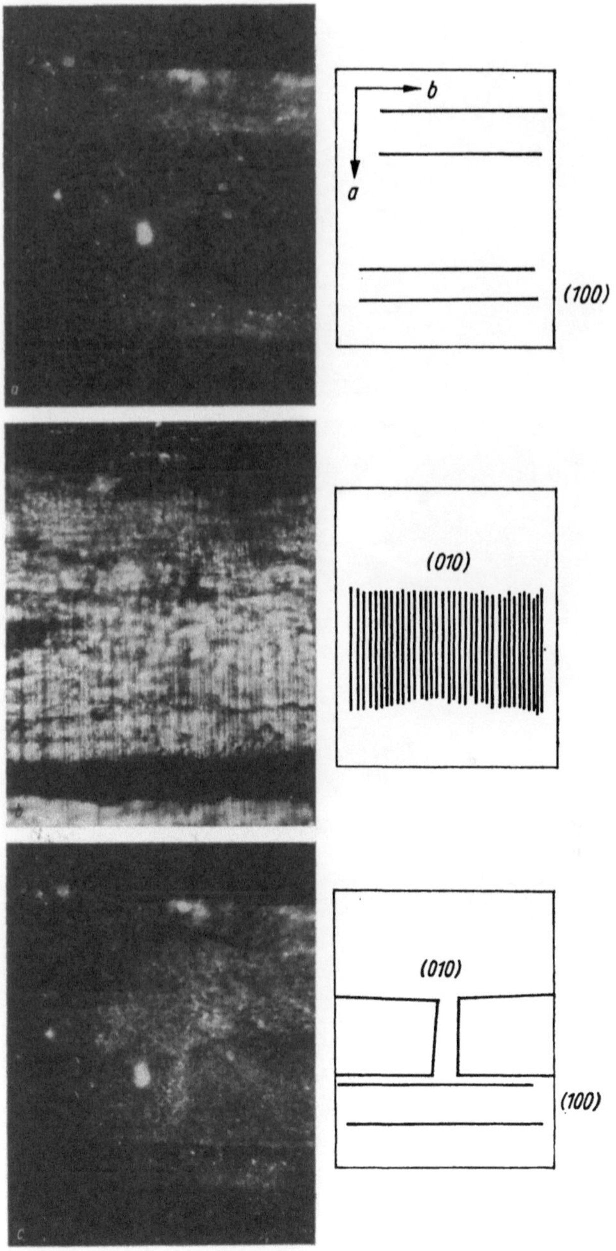

Fig. 4. Domain structure in the c-direction in the crystal vs the temperature: (a) (100) domain walls at room temperature; (b) (010) domains below T_2; (c) (100) and (010) domains observed when the crystal was cooled back to room temperature after heating above T_1

A sample viewed along the **b**-axis at room temperature is shown in Figure 5a. No domain structure is observable except of the twin walls of the type (101) which remain also after heating above T_1. When the sample is cooled down back to room temperature the domain picture becomes more complicated because the large (101) domains have fine subdomain structure with the (001) twin walls (Fig. 5b). Further cooling down below T_2 causes the domains, except of the type (101), to disappear (Fig. 5c). The domain walls (101) and (10$\bar{1}$) observable in all three phases, were regarded as a growth domains rather, since they are temperature independent.

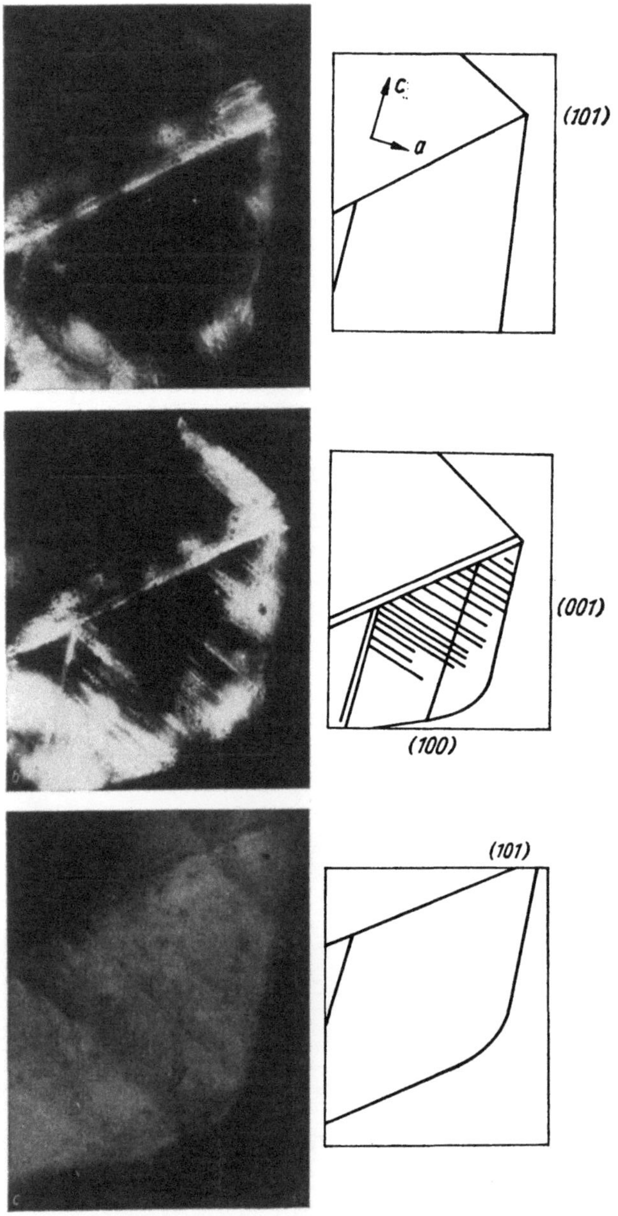

Fig. 5. Domain walls observed in the b-direction: (a) (101) and (10$\bar{1}$) growth domains; (b) (001) domains at room temperature after heating above T_1; (c) below T_2

These results confirm the conclusion drawn from the thermal expansion measurements that at room temperature the crystal is triclinic and monoclinic below T_2.

In order to determine the symmetry of particular phases the X-ray Weissenberg photographs have been taken in all phases. In connection with the former results we can propose the following phase transitions scheme:

$$\text{P11b (III)} \underset{}{\overset{245\,\text{K}}{\rightleftarrows}} \text{P1 (II)} \overset{330\,\text{K}}{\rightleftarrows} \text{Pmnb (I)}\,.$$

The sequence of the phases is fully reversible on the heating and cooling runs.

Table 2
Phase transition sequences in NH_4BeF_3 and ND_4BeF_3 crystals

NH_4BeF_3		ND_4BeF_3 [c]
I-type [a]	II-type [b]	
I (Pmnb)	I (Pmnb)	I (Pmnb)
347 K ↑		
II ($P2_12_12_1$)	334 K ⇅	330 K ⇅
334 K ↑		
III (P1n1) ← (cooling from I)	II (P1)	II (P1)
253 K ⇅	253 K ⇅	245 K ⇅
IV (P1)	III (P1) or monoclinic	III (P11b)

References: a) ŁUKASZEWICZ, WAŚKOWSKA et al.; b) YOSHIDA, TSUKAMOTO et al. and YOSHIDA, TAKEMASA et al.; c) present paper.

In Table 2 there are set together the phase transition schemes in NH_4BeF_3 and ND_4BeF_3 crystals. In NH_4BeF_3 of the first type there are recorded four phases on the heating run with one additional orthorhombic phase II ($P2_12_12_1$) at 334 K, whereas the transition to the Phase I is shifted to the 347 K. The intermediate Phase II is missed on cooling run. Such a heating and cooling cycle is fully repeatable. NH_4BeF_3 crystal of the II-nd type exhibits two only phase transition points. According to the papers of YOSHIDA et al. (1984 a, b) the crystal is triclinic at room temperature and monoclinic or triclinic in the low temperature phase. Similar phase situation is observed in deuterated ND_4BeF_3, where a shift of the transition temperatures is connected with an isotropic effect but the corresponding phases are most likely to be isomorphous.

Discussion

Presented experiments led us to the following statements:

1. ND_4BeF_3 has only two phase transition points in the temperature region between 215–360 K.

2. At room temperature the crystal is triclinic with the space group P1. At 330 K its symmetry changes to orthorhombic. Another Ī-st order phase transition takes place at 245 K where the crystal symmetry changes to P11b or P112/b with the monoclinic axial angle $\gamma \neq 90°$. One can notice that at the room temperature point all the axial angles of the unit cell became equal to $90 \pm 0.02°$ although below and above this temperature within the limits of the triclinic Phase II, the β and γ angles deviate considerably from this peculiar value.

3. At room temperature the crystal exhibits domain structure of the type (100), (010), and (001) disappearing above T_1 and of the type (010) observable below T_2.

4. The whole population of ND_4BeF_3 crystals — studied by us — belonged to the second-type of crystals mentioned earlier. Present results are similar to those reported by YOSHIDA et al. (1984 a, b), describing the phase situation in NH_4BeF_3, with the difference noticed for the c-parameter run in the low temperature phase. In our

ND_4BeF_3 the *c*-parameter considerably elongates at the transition below T_2 whereas in NH_4BeF_3 studied by these authors the lattice constant *c* diminishes in the same temperature region. This effect has connection with a run of the unit cell volume, since the *c* parameter behaviour has a predominant contribution to the volume thermal expansion at the Phase III.

The differences between the two types of crystals may be considered in terms of real structure i.e. differences in the crystal perfection and in distribution of defects what influence a dynamics of the domain formation.

The authors greatly thank Professor K. Łukaszewicz for valuable suggestions and discussions and Dr. T. Niklewski from Arrhenius Laboratory, Stockholm, for the DSC measurements.

References

BOND, W. L.: Acta Cryst. **13**, 814 (1960)
CZAPLA, Z., CZUPIŃSKI, O., WAŚKOWSKA, A.: J. Phys. Soc. Jpn **51**, 2693 (1982)
ŁUKASZEWICZ, K., KUCHARCZYK, D., MALINOWSKI, M., PIETRASZKO, A.: Kristall und Technik **13**, 561 (1978)
ŁUKASZEWICZ, K., WAŚKOWSKA, A., TOMASZEWSKI, P. E., CZAPLA, Z.: Phase Transitions **4**, 47 (1983)
MAKITA, Y., SUZUKI, S.: J. Phys. Soc. Jpn **48**, 693 (1980)
MILKOVA, L. P., PORAY-KOSHITS, M. A.: Izv. Akad. Nauk SSSR **26**, 368 (1962)
TEDENAC, J. C., COT, L.: C.R. Acad. Sci. **268C**, 1687 (1969)
YOSHIDA, H., TAKEMASA, H., OSHINO, Y., MAKITA, Y.: J. Phys. Soc. Jpn **53**, 2600 (1984)a
YOSHIDA, H., TSUKAMOTO, T., FUTAMA, H., MAKITA, Y.: J. Phys. Soc. Jpn **53**, 2606 (1984)b
WAŚKOWSKA, A.: Acta Cryst. **C39**, 1167 (1983)
WAŚKOWSKA, A.: Acta Cryst. **C41**, 1714 (1985)

(Received December 29, 1986)

Authors' addresses:

Alicja WAŚKOWSKA, Adam PIETRASZKO
Institute for Low Temperature and Structure Research, Polish Academy of Sciences,
pl. Katedralny 1
50950 Wrocław, Poland
Zbigniew CZAPLA
Institute of Experimental Physics, University of Wrocław,
ul. Cybulskiego 36
50205 Wrocław, Poland

Cryst. Res. Technol.	22	1987	7	951–959

J. HÄRTWIG, K.-H. JÄCKEL, V. LERCHE

Friedrich-Schiller-Universität Jena, Sektion Physik, Jena, and
Akademie der Wissenschaften der DDR, Zentralinstitut
für Physik der Erde, Institutsteil Jena

Simulation of Contrasts in X-ray Plane Wave Topographs of Quartz Crystals with Induced Growth Striations

Using system theoretical methods and fast discrete Fourier transforms, contrasts in plane wave topographs of crystals with impurity distributions of any form can be simulated with high accuracy and in a short time.

Bei Anwendung von übertragungstheoretischen Methoden und schnellen diskreten Fourier Transformationen können Kontraste in Planwellentopogrammen mit Fremdstoffverteilungen von beliebiger Form mit großer Genauigkeit und in kurzer Zeit simuliert werden.

1. Introduction

Recently the method of induced growth striations (DAMEN, ROBERTSON; GÖRNERT, HERGT) has been successfully applied also for the hydrothermal growth of quartz crystals (BERNHARDT, ALTER). By means of this method it is possible to study and to understand the local growth behaviour of quartz and to obtain a deeper insight in the impurity atom incorporation in the crystal in dependence of the growth parameters. This is of great importance, because a direct observation of the local growth process is impossible in the case of quartz. Quartz crystals with induced growth striations may be investigated by means of optical (BERNHARDT, ALTER) and X-ray topographical (ALTER, HÄRTWIG; ALTER et al. 1987) methods. Especially in the case of very low changes of the impurity atom/ion concentration, which are the cause of the mentioned growth striations, X-ray plane wave topography is an appropriate experimental method to detect the week lattice distortions of the striations. However, to use the full information content of the topographs in an optimum way one must have available efficient methods for the simulation and analysis of topographs.

ALTER et al. (1987) presented a qualitative and quantitative study of the contrast formation in X-ray plane wave topographs of quartz plates with induced growth striations. The aim of the present paper is to improve the mathematical treatment of such kind of problems by a new approach.

This new approach consists of two parts:

1. A system theoretical description of both, the dynamical response of the autoclave to an outer control parameter, and the elastic response of the crystal lattice of a crystal plate to the impurity distribution,
2. The application of a fast discrete Fourier transformation. The advantages of these improvements are:

— a simple treatment of any form of the temporal dependence on the outer control parameter,

— possibilities of taking into account a non-constant local growth velocity,
— a considerable decreases of computational time,
— a higher quality of the numerical Fourier transformation (no effects of truncation errors).

This new approach also ought to make possible computer simulation of the X-ray contrast according to the Takagi theory within reasonable computational times and also the treatment of the crystal deformation according to the anisotropic theory of elasticity.

2. Theoretical description

Similar like ALTER et al. (1987) we shall divide the entire problem into three parts.

2.1. Connection between a modulated outer control parameter and the local impurity distribution in the crystal

A linear time-invariant system with the input parameter $u(t)$ and the output parameter $y(t)$ may be fully characterized by a transfer function $H(\omega)$. If Y and U are the Fourier transforms of y and u, defined for example by $y(t) = \int_{-\infty}^{\infty} Y(\omega) \exp(i\omega t)\, d\omega$, then the relation $Y(\omega) = H(\omega) \cdot U(\omega)$ holds in the frequency space. This may be applied also to the system "autoclave", and we can look for its transfer function.

Let us assume that the connection of an outer control parameter $\delta p(t)$ (modulation of the difference of the applied heating power $\Delta p(t) = \Delta p_0 + \delta p(t)$ between nutrient dissolving zone and crystal growing zone in the autoclave) or $\delta T_s(t)$ (modulation of the to $\Delta p(t)$ related temperature difference $\Delta T_s(t) = \Delta T_{s0} + \delta T_s(t)$) and the temperature response within the growth vessel $\Delta T(t) = \Delta T_0 + \delta T(t)$ may be described by the following differential equation of first order

$$\tau \frac{d(\delta T(t))}{dt} + \delta T(t) = cp(t) = \delta T_s(t) \tag{1}$$

where τ is the time constant of the system and c is a constant parameter.

Carrying out the Fourier transformation we obtain the transfer function $H_A(\omega)$ of the autoclave

$$H_A(\omega) = (1 + i\omega\tau)^{-1}\,. \tag{2}$$

Solving equ. (1) using equ. (2) and the related back transformation we obtain a result identical with that of ALTER et al. (1987).

We assume that the temporal dependence of the amount of the impurity incorporation in the crystal $\tilde{c}(t) = c_0 + \delta\tilde{c}(t)$ is the same like the temperature variation $\Delta T(t)$, so that $\tilde{c}(t) = b \cdot \Delta T(t)$. Here b is a constant and c_0 the mean concentration of impurities in the crystal. The functional dependence of the local distribution of impurities in the crystal as a function of the position $c(x) = c_0 + c\delta(x)$ is assumed to be similar to the temperature variation as a function of the time. But it must not be exactly the same, because in general the growth velocity $v_w(\Delta T(t)) = dx/dt$ is not constant, and thus there is no linear relation between the time t and the position x. As a general relation we can assume that

$$c(x) = c(\Delta T(t(x)))\,. \tag{3}$$

Thus, in principle, it is possible to calculate $c(x)$ from the knowledge of $\Delta T(t)$ and $H_A(\omega)$.

For our practical application not the straight forward calculation of $c(x)$ is of interest, but the determination of the local growth velocity. This is possible comparing $\tilde{c}(t)$ and $c(x)$, where the last distribution is obtained from experimental results by analysing X-ray topographs.

2.2. Connection between local impurity distribution in the crystal and the local strain fields of the crystal plate

The deformation of a crystal plate with an impurity distribution $c(x)$ that takes into account the surface strains caused by the relaxation of internal concentration stresses was calculated by HÄRTWIG, and by ALTER on the basis of results of TIMOSHENKO, GOODIER, and of INDENBOM, KROUPA. This theoretical treatment was carried out according to the isotropic theory of elasticity. Its results may be easily translated into the language of transfer theory. With this approach also a relatively simple extension to the anisotropic theory of elasticity is possible. A paper with a detailed treatment of these problems will be prepared in near future. Here only some results of the transfer theoretical approach for the isotropic theory of elasticity shall be reported.

Let us introduce a coordinate system in such a way that its origin is situated on the crystal surface. The x-axis is parallel to the crystal plate surface and perpendicular to the induced growth striations. The y-axis is perpendicular to the surface of the crystal plate and parallel to the growth striations (compare HÄRTWIG or ALTER et al. 1987). The change of the concentration distribution $\delta c(x)$ leads directly to a distribution of the relative lattice parameter change of the infinite crystal (in the growth direction)

$$\frac{\Delta d_x^\infty}{d_x}(x) = \frac{1}{2}\left[\max\left(\frac{\Delta d_x^\infty}{d_x}\right) - \min\left(\frac{\Delta d_x^\infty}{d_x}\right)\right] f(x) = kf(x) = \frac{1+\nu}{1-\nu}\, k_{\mathrm{m}} f(x)\,. \tag{4}$$

Here we introduced a function $f(x)$ proportional to $\delta c(x)$ but normalized in such a way that $-1 \leqq ||f(x)|| \leqq 1$. ν is the Poisson ratio. The following fields are of interest for calculating X-ray contrast in the symmetrical Bragg case (HÄRTWIG):

a) The total relative change of the lattice parameter d_y^T (perpendicular to the crystal surface) within the crystal plate:

$$\frac{\Delta d_y^T}{d_y}(x, y) = e_{yy}^T(x, y)\,. \tag{5}$$

b) The tilt angle γ_x of the lattice planes which were parallel to the x–z plane before deformation:

$$\gamma_x(x, y) = \frac{\partial}{\partial x}\int e_{yy}^T(x, y)\, dy\,. \tag{6}$$

Here e_{yy}^T is a component of the strain tensor. If $F(v)$, $E_{yy}^T(v, y)$, $\Gamma_x(v, y)$ are the Fourier transforms of $f(x)$ $e_{yy}^T(x, y)$, $\gamma_x(x, y)$ with respect to the x-coordinate then we can write

$$E_{yy}^T(v, y) = H_{yy}^T(v, y) \cdot F(v)\,, \tag{7}$$

$$\Gamma_x(v, y) = H_{\gamma x}^T(v, y) \cdot F(v)\,. \tag{8}$$

H_{yy}^T and $H_{\gamma x}^T$ are transfer functions of a crystal plate with respect to the total relative change of the lattice parameter d_y^T and the tilt angle γ_x. They only depend on crystal properties (thickness, Poisson ratio) and not on the concentration distribution. The latter enters theory by a simple multiplication by its Fourier transform $F(v)$. Thus, once the functions H_{yy}^T and $H_{\gamma x}^T$ were calculated we can obtain the results for $\Delta d_y^T/d_y$ and

γ_x for every impurity distribution (if it has a Fourier transform). The upper results are in accordance with those of HÄRTWIG, if we put

$$H_{yy}^T(v, y) = kI_2(v, y) , \tag{9}$$

$$H_{\gamma x}^T(v, y) = ikI_6(v, y) . \tag{10}$$

Thus it is possible also to treat the deformation of a crystal plate within the transfer theoretical framework and to use its advantages.

2.3. Connection between local strains and optical density in an X-ray plane wave topograph

To calculate the distribution of the optical density in a topograph we used (similar like ALTER et al. 1985, 1987) the approximation of the local application of the dynamical theory for plane waves and a perfect crystal. X-ray contrast in the symmetrical Bragg case depends on the following quantity (BONSE):

$$\Delta s(x, y) = \boldsymbol{n}_h \cdot \boldsymbol{n}_t \gamma_x + \tan \theta_B \, \Delta d_y^T / d_y . \tag{11}$$

Here $\boldsymbol{n}_h$, $\boldsymbol{n}_t$ are unit vectors parallel to the diffractometer axis and the axis of the lattice tilt, respectively. θ_B is the Bragg angle.

Knowing the form of the rocking curve of the used experimental arrangement and the distribution of Δs near the crystal surface it is possible to calculate the locally reflected normalized intensity $I_n = I/I_0$ in the point x. Taking into account the curve of density we can convert the normalized intensity distribution into a distribution of the optical density $S(x)$ in an X-ray topograph.

Thus we found a theoretical description for the entire problem starting with $\Delta T_s(t)$ and finishing with $S(x)$.

3. Calculations

The first step in the realisation of our theoretical model is the calculation of the transfer function $H_A(\omega)$ of the autoclave. Figure 1 shows the shapes of its real and imaginary parts for two time constants τ. For an outer control parameter $\delta T_s(t)$ as shown in Figure 2 (taken from the experiment) we calculated $\delta T(t)$ by means of the

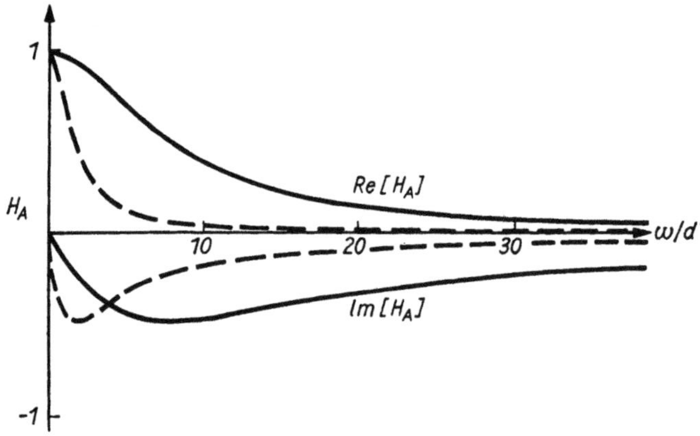

Fig. 1. Real and imaginary parts of the transfer function $H_A(\omega)$ of the autoclave for two time constants τ. Straight line: $\tau = 0.125$ d; broken line: $\tau = 0.5$ d

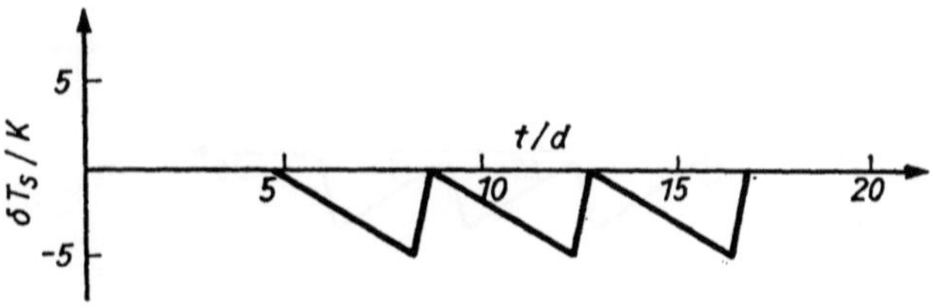

Fig. 2. Practical example of a shape of the control parameter $\delta T_s(t)$

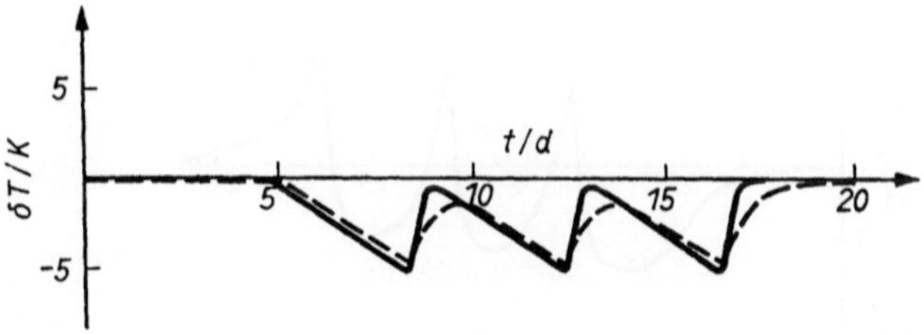

Fig. 3. Shapes of $\delta T(t)$ calculated by means of the two transfer functions shown in Figure 1. Straight line: $\tau = 0.125$ d; broken line: $\tau = 0.5$ d

transfer functions for two values of τ (Fig. 3). Characteristical changes of the shape of $\delta T(t)$ in dependence of τ are visible. The numerical value of τ may be estimated comparing experimental and calculated results of intensity distributions in topographs.

The shapes of the normalized transfer function of the crystal plate (thickness $D = 1$ mm) with respect to the local relative change of the lattice parameter d_y^T and the tilt angle γ_x H_{yy}^T and $H_{\gamma_x}^T$ are shown in Figure 4 for $y = 0$ and a Poisson ratio $\nu = 0.057$. By means of $H_{yy}^T(v, y)$, $H_{\gamma_x}^T(v, y)$ and $F(v)$ it is possible to calculate $E_{yy}^T(v, y)$ and $\Gamma_x(v, y)$ in a straight forward way and with that the fields $e_{yy}^T(x, y)$ and $\gamma_x(x, y)$. For the case of constant growth velocity v_w the function $F(v)$ may by calculated directly from the product of the Fourier transform of $\delta T_s(t)$ and $H_A(\omega)$ by a simple scale transformation. The results for $\gamma_x(x, 0)$ and $e_{yy}^T(x, 0)$ are shown in Figure 5 and (because of the fact that $\tan \theta_B \approx 1$ for the used Bragg angle $\theta_B = 46.4°$) in Figure 6a, respectively. The shapes of $\delta T(t)$ (Fig. 3) and $e_{yy}^T(x, 0)$ (Fig. 6a) are also very similar because $H_{yy}^T(v)$ is nearly constant and thus practically only $F(v)$, which has for constant v_w the functional form of the Fourier transform of $\delta T(t)$ (see above), influences the shape of e_{yy}^T.

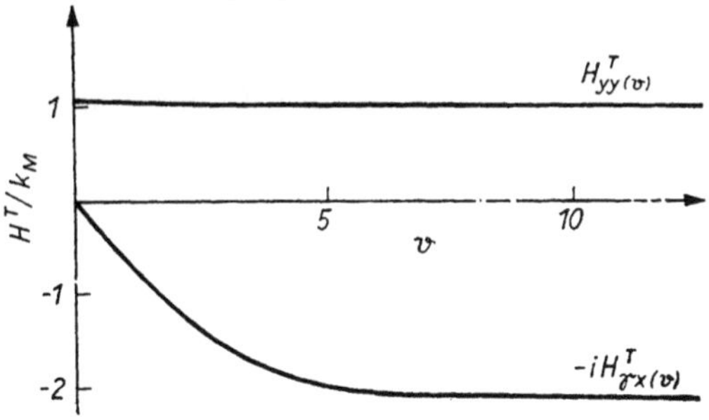

Fig. 4. Shapes of the transfer functions $H_{yy}^T(v)$ and $H_{\gamma_x}^T(v)$ normalized with respect to k_m. $D = 1$ mm, $y = 0$, $\nu = 0.057$

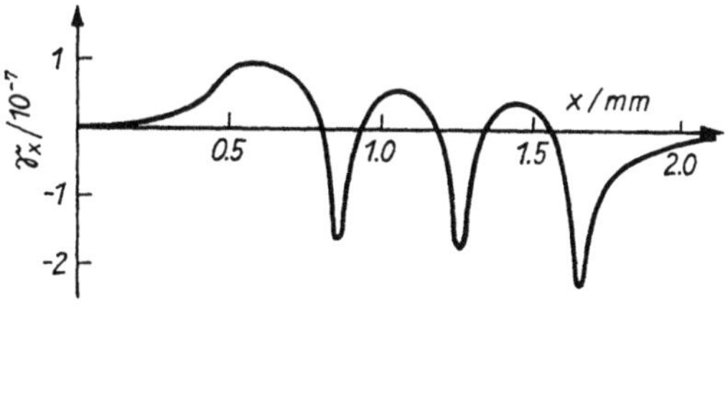

Fig. 5. Shape of $\gamma_x(x, 0)$ calculated by means of the tranfser function $H_{\gamma_x}^T(v)$ and $H_A(\omega)$ for $\tau = 0.125$ d, $v_w = 0.1$ mm/d, $k_m = 10^{-7}$, $D = 1$ mm

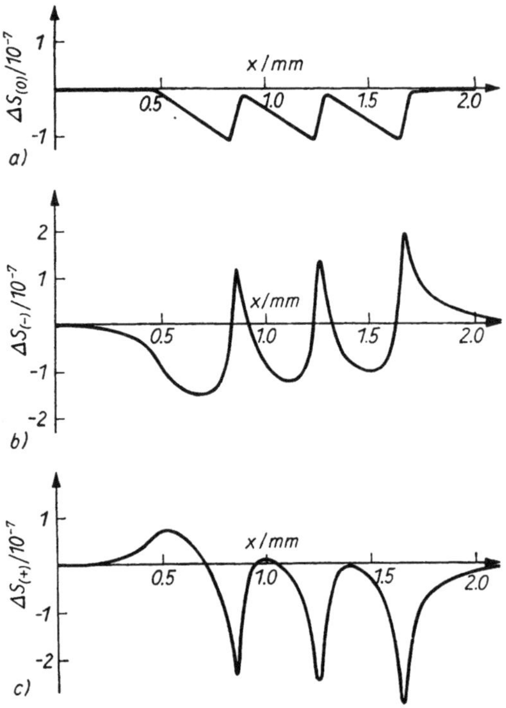

Fig. 6. Shapes of Δs for $n_h \cdot n_t = 0$ (a), $n_h \cdot n_t = -1$ (b), $n_h \cdot n_t = +1$ (c). $\tau = 0.125$ d, $v_w = 0.1$ mm/d, $k_m = 10^{-7}$, $D = 1$ mm, $y = 0$

Distributions of the quantities $\Delta s_{(0)}$, $\Delta s_{(-)}$, $\Delta s_{(+)}$ at the crystals surface for three different values of the product $n_h \cdot n_t = 0, -1, +1$ are shown in Figure 6 a—c.

Like mentioned in section 2.3. the intensity reflected from the crystal depends of Δs. Figures 7 b, d and 8 b, d show normalized intensity distributions calculated for different values of $n_h \cdot n_t$, for working point on different sides of the rocking curve (but always on a position where it is most steeply) and for two kinds of the form of the outer control parameter $\delta T_s(t)$.

4. Comparison with experimental results

In Figures 7 and 8 measured normalized density distributions are compared with normalized intensity distributions. A good qualitative accordance can be stated. Figures 9 a, b show the plane wave topographs from which the photographic density distribution were taken. A specific property of the contrast distribution in the cases $n_h \cdot n_t = \pm 1$ (Fig. 8) is an increase or decrease of the maximum or minimum values of the modulation within one group of modulations. For the description of thuch distributions we must work with finite numbers of modulations, as in reality. ALTER et al. (1987) used periodical functions $\delta T_s(t)$, which describe the temperature modulation from minus to plus infinity. In that case the shape of the density distribution of one period is reproduced quite well, but without the mentioned increase or decrease within one group of modulations. The above described system theoretical approach makes it possible to overcome those limitiations. Distinct differences between theoretical and experimental result appear at position where $\delta T_s(t)$ changes very fast. So the spikes visible in the theoretical distributions do not appear in the experimental ones. There are several reasons for these differences: 1. The approximation of the local

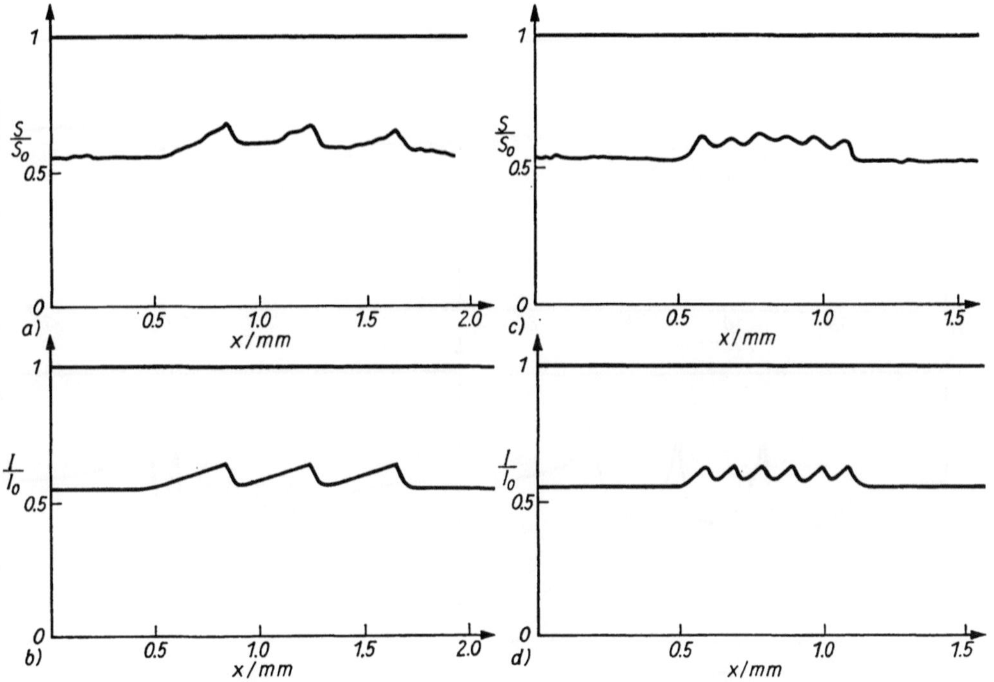

Fig. 7. Comparison of measured normalized photographic density distributions (a, c) with normalized intensity distributions (b, d) calculated for a working point at 55 % of the maximum intensity on the high angle side of the rocking curve. $n_h \cdot n_t = 0$, $\tau = 0.125$ d, $v_w = 0.1$ mm/d, $k_m = 10^{-7}$. In a and b the outer control parameter $\delta T_s(t)$ exhibited three periods of about four days duration (as shown in Figure 1) and in c and d six periods of about one day

applicability of the dynamical theory for plane waves and a perfect crystal is not a good one in these places, 2. the application of the isotropic theory of elasticity, 3. the calculation with Δs at the crystal surface only, 4. the calculation with a constant growth velocity (computer experiments showed that this approximation is of less importance), 5. the not ideal resolving power of the double crystal spectrometer (for a distance between crystal and photographic plate of 3 cm the picture of a point has a diameter of about 10 μm; the distance between the modulations in Figure 9a is about 70 μm).

Because the shape of $\delta T(t)$ (Fig. 3) and $\Delta s_{(0)} = \tan \theta_B \cdot e_{yy}^T$ (Fig. 6a) are nearly identical, it is possible to estimate the parameter τ from the comparison of measured normalized photographic density distributions with normalized intensity distributions in the case $n_h \cdot n_t = 0$ (Fig. 7). The ratio of the distance between the maximum and the minimum of one period of modulations and of the distance between two subsequent maxima or minima (length of the period) sensitively depends on the parameter τ. By fitting the theoretical to the experimental ratios values $\tau = (0.10 \ldots 0.20)$ d could be obtained. Thus, the value $\tau = 0.125$ d used for the simulations is in good agreement with the experiments. It also agrees quite well with the value $B = (\tau v_w)^{-1} = 0.09\ \mu m^{-1}$ (ALTER et al. (1987) for $\tau = 0.11$ d and $v_w = 0.1$ mm/d.

The parameter k_m may be also estimated by fitting the theoretical intensity distributions to the experimental density distributions. A critical point is here the exact knowledge of the working point. A good agreement could be obtained for $k_m =$

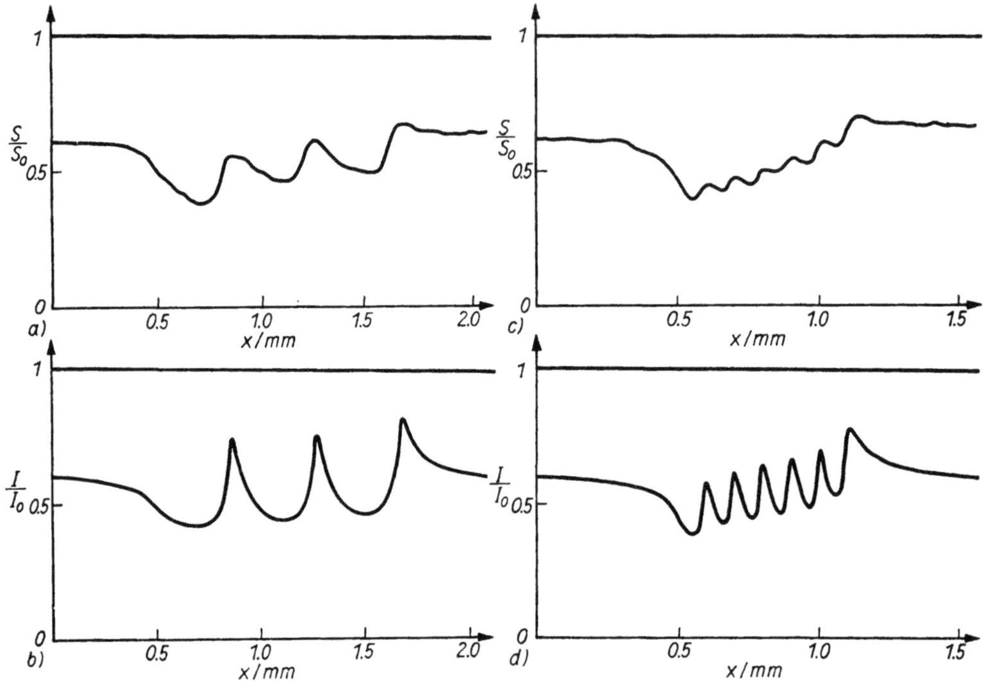

Fig. 8. Comparison of measured normalized density distributions (a, c) with normalized intensity distributions (b, d) calculated for a working point at 60 % of the maximum intensity on the low angle side of the rocking curve and for the same shapes of the outer control parameter as in Figure 7

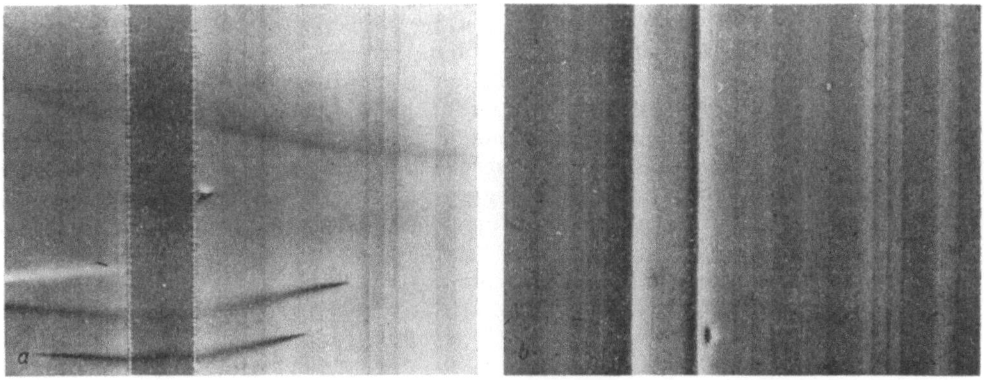

Fig. 9. Plane wave topographs from which the photographic density distributions 7 a, c and 8 a, c were measured (Fig. 7 a, c belongs to Fig. 9a, Fig. 8 a, c to Fig. 9b)

$= (1.00 \pm 0.25) \cdot 10^{-7}$ for the last two modulation groups in Figure 1 of ALTER and HÄRTWIG and for the half of this value for the first two modulation groups.

To calculate the mean local growth velocity $\bar{v}_w$ we must measure the distances $\Delta L(x)$ between distinct points in the density distribution and the (temporal) distances Δt between distinct points in the outer control parameter $\Delta T_s(t)$. Then holds $\bar{v}_w(x) = \Delta L(x)/\Delta t$. In the case of small concentrations where the working point shifts only a little on one side of the rocking curve the most suited points for the determination of $\Delta L(x)$ are the spikes (or what had to be a spike) in the intensity distributions for the cases $\boldsymbol{n}_h \cdot \boldsymbol{n}_t = \pm 1$. The position of these maxima is nearly independant of model parameters like τ and especially nearly independant of v_w for its own.

5. Conclusions

The new mathematical approach for the simulation of contrasts in plane topographs of crystals with inhomogeneous distributions of impurities provides a good accordance of theoretical and experimental results. The results could be obtained on a micro(home)-computer. Comparing experimental and theoretical results the values of several model parameters could be determined.

References

ALTER, U., CUAMBA, B., HÄRTWIG, J.: Cryst. Res. Technol. **22**, 217 (1987)
ALTER, U., HÄRTWIG, J.: Cryst. Res. Technol. **20**, 1173 (1985)
ALTER, U., HÄRTWIG, J., KUBĚNA, J.: Czech. J. Phys. B **35**, 158 (1985)
BERNHARDT, Hj., ALTER, U.: Cryst. Res. Technol. **19**, 453 (1984)
BONSE, U.: Direct Observation of Imperfections in Crystals. Interscience, New York—London, 1962, p. 431
DAMEN, J. P. M., ROBERTSON, L. M.: J. Cryst. Growth **16**, 50 (1972)
GÖRNERT, P., HERGT, R.: phys. stat. sol. (a) **20**, 577 (1973)
HÄRTWIG, J.: Cryst. Res. Technol. **16**, 1297 (1981)
INDENBOM, V. L., KROUPA, F.: Czech. J. Phys. B **29**, 333 (1979)
TIMOSHENKO, S., GOODIER, I. N.: Theory of Elasticity, New York, Toronto, London 1951

(Received December 22, 1986)

Authors' addresses:

Dr. J. HÄRTWIG, V. LERCHE
Friedrich-Schiller-Universität Jena, Sektion Physik
Max-Wien-Platz 1
Jena
DDR-6900

K.-H. JÄCKEL
Akademie der Wissenschaften der DDR,
Zentralinstitut für Physik der Erde, Institutsteil Jena
Burgweg 11
Jena
DDR-6900

Cryst. Res. Technol.	22	1987	7	961–967

G. Bochmann, E. Klaus, G. Fritsche, W. Wagner

Technische Universität Karl-Marx-Stadt, Sektion Chemie und Werkstofftechnik

Phase Structure of Boride Layers on Iron Substrates with Electroplated Cobalt Layers

Specimens made of Armco iron were electroplated with cobalt and then boronized with 2.2 vol% BCl_3 in H_2 at 1223 K. The boride layers obtained show the sequence of crystallographic phases which is (Co, Fe)B, FeB, Fe_2B, substrate material.

Proben aus Armco-Fe wurden mit einer galvanischen Kobaltschicht versehen und mit 2.2 Vol% BCl_3 in H_2 bei 1223 K boriert. Die dabei entstehenden Boridschichten weisen die Phasenfolge (Co, Fe)B, FeB, Fe_2B, Substratwerkstoff auf.

1. Introduction

Tools and components made of metallic materials may be coated with boride layers to improve their wear resistance (Fichtl; Heydecker; Zoellner; Riehle). Boronizing is one of several methods of thermochemical treatment. The most widely used mixture producing boride layers in the gas phase is trichlorborane (BCl_3) and hydrogen (Bochmann et al. 1982). The boride layers formed on iron substrates may consist of the two phases FeB and Fe_2B. The formation of a toothlike structure causes a good adherence of the boride layer to the substrate. If more transition metals, apart from iron, exist in the surface layer of a specimen these are apt to influence the phase structure of the boride layer. Thus, it was proved that if nickel layers were electroplated on iron substrates and annealed the phase sequence caused by subsequent boronizing was $(Ni, Fe)_2B$, FeB, Fe_2B as seen from outside in an inward direction (Bochmann et al. 1984). Obviously, the reason is that there is isomorphism between Ni_2B and Fe_2B (tetragonal, I4/mcm) and that there are differences in crystal structure between NiB (orthorhombic, Cmcm (Wyckoff)) and FeB (orthorhombic, Pnma (Wyckoff)). Now, when a metall is present in the surface layer where the two borides, i.e. MeB and Me_2B, are isomorphous with the iron borides, the phase sequence that could be expected should again be MeB, Me_2B, substrate. To verify this, iron specimens were electroplated with cobalt coatings, boronized in the gas phase, and analyzed to prepare the present paper.

The main objective of diffusing more transition metals into the iron boride layer is virtually to modify its properties, e.g. surface morphology, corrosion resistance, adhesion, and so on.

2. Experimental

The laboratory boronizing outfit was described earlier (Bochmann et al. 1982). The Armco iron specimens (MK3Al, TGL 101–18499) with dimensions 0.6 cm × 0.6 cm × 1.0 cm were wetground to 500 mesh, degreased, anodically roughned (60% H_2SO_4), and electroplated at room temperature with a cobalt layer (bath composition $CoSO_4 \cdot 7\ H_2O$ 504, NaCl 17, H_3BO_3 45 g/l, $pH = 5$, current density = $4\ A \cdot dm^{-2}$ (DETTNER, ELZE)). These specimens were annealed in H_2 at 1223 K for one hour and boronized immediately afterwards with 2.2% by volume BCl_3 in H_2 (5 L/h) for 2.5 hours. X-ray diffraction analysis was performed by Co Kα radiation. Scanning electron microscopy (SEM) pictures were made by means of a JSM-U3 Jeol, Japan. Cross-sections of the specimens were produced after electroplating with a nickel layer about 5 µm in thickness (Watt's bath (DETTNER, ELZE)) by wet grinding to 500 mesh and polishing with diamond pastes to measure the thickness of the boride layers and to carry out electron probe X-ray microanalysis (EPXMA) investigations (using the JXA equipment, Jeol, Japan).

3. Results and discussion

Cobalt will diffuse into the substrate and iron into the electroplated layer which rest evenly on the Armco iron (Fig. 1a) when the specimens are subjected to annealing. Whatever the initial thickness might be (5; 10; 20 µm) cobalt will diffuse into the bulk material up to a depth of about 10 µm within one hour at 1223 K. According EPXMA investigations the maximum concentration of cobalt on the surfaces of the specimens is about 50% by weight (5 µm Co) and around 90% by weight (10; 20 µm Co) after annealing procedure mentioned above.

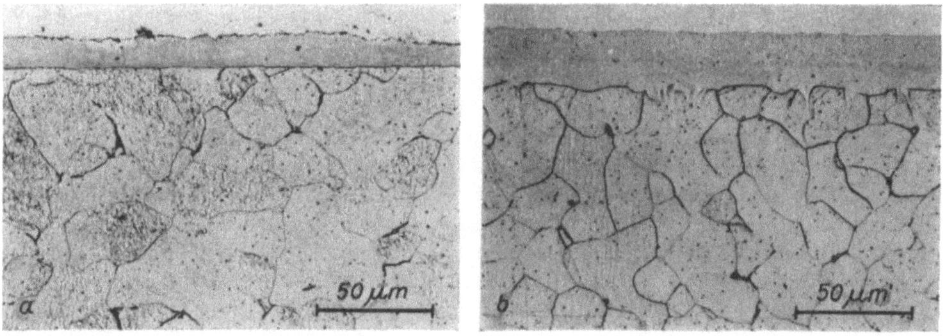

Fig. 1. Optical micrograph of an electroplated cobalt layer on Armco iron (MK3Al), a) untreated, b) after annealing at 1223 K for 4 hours

Boronizing causes the specimens to gain in weight which amounts to $8.8 \pm \pm 0.5$ mg/cm² after a boronizing time of 2.5 hours at 1223 K (2.2 vol% of BCl_3 in H_2). A smaller amount (7.4 ± 0.2 mg/cm²) for the specimens having the cobalt layer 5 µm in initial thickness may proceed from a stronger effect of the corrosive attack originated by the boronizing gas stream because of the increased iron content in the surface layer (BOCHMANN et al. 1984). A layer rich in cobalt may relatively well protect the substrate against the corrosive attack, and may give rise to improved adhesion of the boride layer (CARLSSON) and a surface morphology of sufficient quality as well. Thus, the surface morphology of the specimen shown in Figure 2 is very regular as it consists of tiny particles as in the case of boronized surfaces protected by a sufficiently large nickel component (BOCHMANN et al. 1984). On the other hand, Armco iron specimens boronized in BCl_3–H_2 mixtures at 1223 K without any specific precaution show larger surface particles (BOCHMANN et al. 1985).

The optical micrographs of the cross-section of the boronized specimens reveal a boride layer composed of two domains and dovetailed typically with the bulk

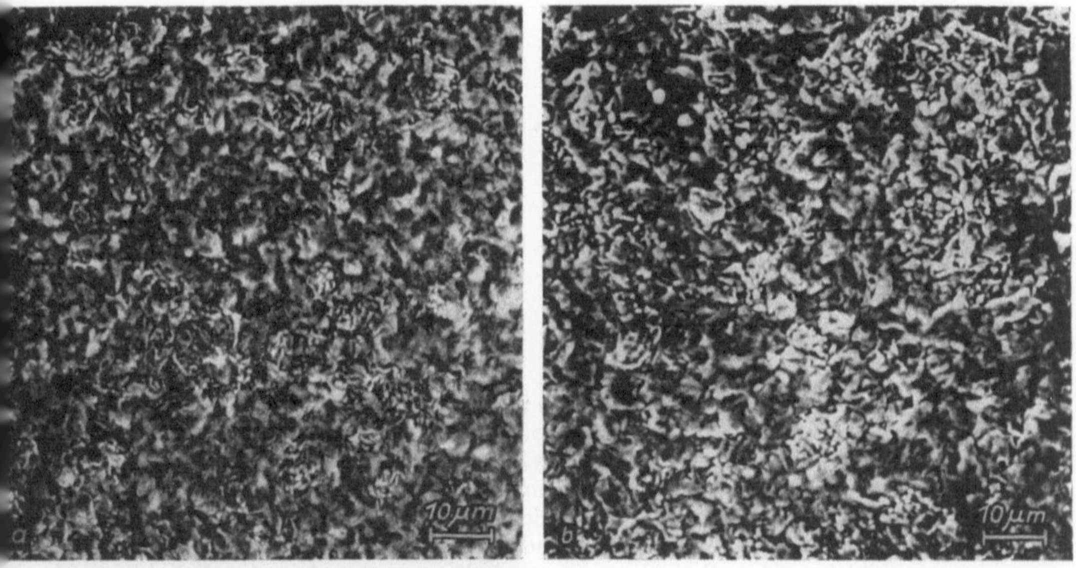

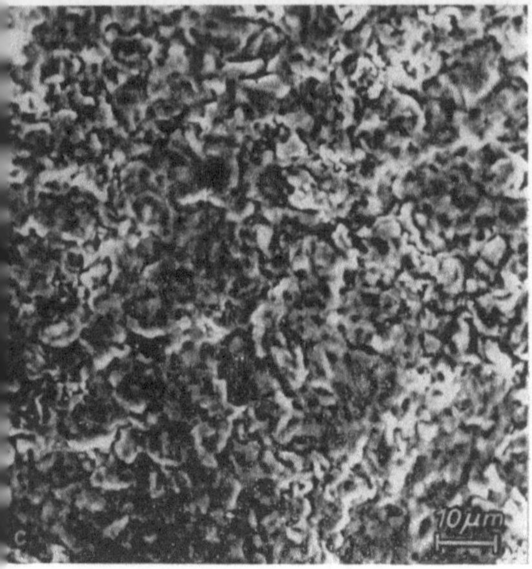

Fig. 2. SEM pictures of an Armco iron specimen electroplated with 10 µm Co and boronized at 1223 K by 2.2 % by volume of BCl_3 in H_2 (5 l/h) for 2.5 hours, distance from the front a) 0.1 cm, b) 0.5 cm, c) 0.9 cm

material (Fig. 3). The full layer thickness measured according TGL 34-993 is about 120 to 130 µm, the thickness of the first domain is about 50 to 60 µm at boronizing conditions given (see Sect. 2).

The micro-hardness values for this outer part of the boride layers, which are $HV_{0.05} = 1100$ to $HV_{0.05} = 1250$, are by far lower than those given for FeB in literature (MATUSCHKA), however, they are within a range indicated for the cobalt borides (LEUSCHEN). As EPXMA investigations point out the outer part contains about 16% per wight boron and the cobalt constituent as well. Hence, it might be termed (Co, Fe)B. The inner portion of the borid layer containing about 8.8% per wight boron can be referred to as Fe_2B. Figure 4 is a diagrammatical view of the courses of the Co and Fe contents as obtained during EPXMA investigations.

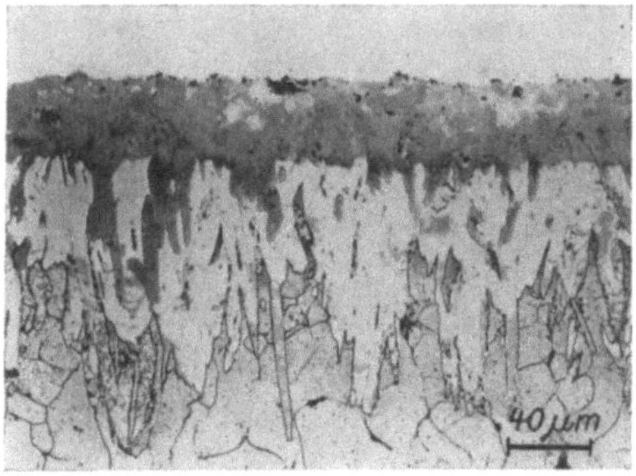

Fig. 3. Optical micrograph of the cross-section of a boride layer obtained on Armco iron electroplated with 20 μm Co (boronizing parameters: 2.2 vol% BCl_3 in H_2 (5 l/h), 1223 K 2.5 hours)

The depths of penetration of X-radiation may be calculated (MIRKIN), the maximum depths of penetration of the CoK_α radiation reaches values between 20 μm to 30 μm for CoB and FeB, respectively (BOCHMANN et al. 1988). Therefore, the X-ray diffraction patterns originate exclusively from the outer crystal phase of the boride layer when the surface of the specimen are subjected to X-ray diffraction analysis. Table 1 lists up a part of the X-ray diffraction found. The specimens were lined up in the table in an order which reflects their decreasing cobalt content and confronted there with a boronized Armco iron specimen. Any variation of the cobalt content in the outer crystal phase produces a continuous alteration of the crystal lattice spacings *d*. There were compared with *d* values (b lines) calculated from lattice constants and Miller's indices for the lattice planes (BOCHMANN et al. 1984, 1988). The lattice constants of the (Co, Fe)B phase with the Co content in sequence 37.5, 25, and 12.5 at% were determined by linear approach between the CoB and FeB pure phase lattice constants listed in the ASTM card index. Table 1 shows that the values of the MeB (111) crystallographic plane are nearly constant, increase with decreasing Co content for the (110), (020), (021) planes, and decrease with dropping Co content for the (101), (002) planes in conformance between the corresponding lines a and b. A comparison between the relative intensities (BOCHMANN et al. 1988) discloses strikingly that the intensified

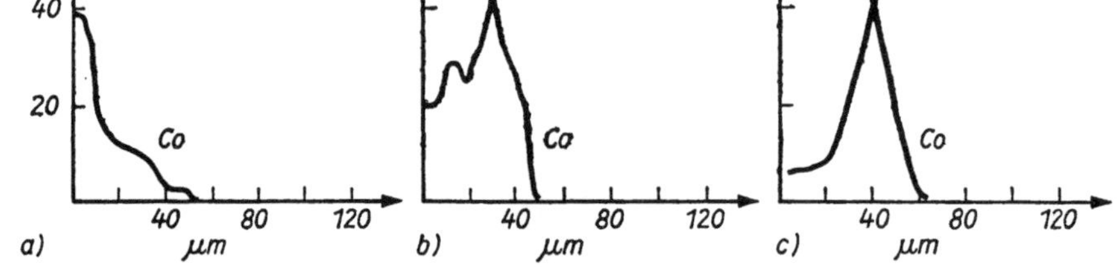

Fig. 4. Turns of Co and Fe contents perpendicular to the surface of the specimens shown diagrammatically according to EPXMA investigations (boronizing parameters as in Figure 2, initial thickness of electroplated Co layer a) 5 μm, b) 10 μm, c) 20 μm)

Table 1

Part of X-ray diffraction analysis (CoK_α radiation) of boronized Armco iron surfaces (boronized with 2.2 vol.% BCl_3 in H_2 for 2.5h at 1223 K), c initial thickness of the electroplated Co layer, d crystal lattice spacing (10^{-10} m, a: measured, b: calculated), I/I_1 relative intensity, (hkl) Miller's indices, sh shoulder, br broad

Assignment		CoB (ASTM)		(Co, Fe)B						FeB	
(hkl)				c: 20 μm		c: 10 μm		c: 5 μm		c: 0 μm	
		d	I/I_1	d	I/I_1	d	I/I_1	d	I/I_1	d	I/I_1
(110)	a	3.14	60	3.187	7	3.202	1	3.236	8	3.285	5
	b	3.154		3.182		3.208		3.235		3.262	
(020)	a	2.62	60	2.648	13	2.661	5	2.720	13	2.767	20
	b	2.622		2.654		2.685		2.717		2.748	
(101)	a	2.41	100	2.415	75	2.408	60	2.408	70	2.398	50
	b	2.407		2.402		2.396		2.390		2.383	
(111)	a	2.19	100	2.196	100	2.196	100	2.196	100	2.199	50
	b	2.188		2.188		2.188		2.188		2.186	
(021)	a	1.99	80	1.995	80	1.983	50	2.008	60	2.018	70
	b	1.984		1.992		1.998		2.004		2.009	
(002)	a	1.52	60	1.518	br 20	1.509	17				
						1.503	15	1.486	br 50	1.477	100
				1.498	sh	1.491	br 13	1.476	br 50		
	b	1.519		1.508		1.496		1.485		1.473	

reflection of the (Co, Fe)B phase can be attributed to the (111) plane as the ASTM card index does for CoB, whereas the most intensified reflection of the FeB phase is called forth by the (002) plane.

An X-ray diffraction diagram from the surface of a boronized specimen whose initial cobalt layer was 20 μm in thickness was produced and another diagram was made for each procedure after the specimen underwent surface grinding in steps of 20, 40, 50, 60, 70, and 80 μm. The comparison of these diagrams proves that the (Co, Fe)B phase is situated in the outermost portion of the boride layer and the earlier discussed variation of d values for various lattice planes hints at the fact that the Co content of (Co, Fe)B diminishes in bulk direction. However, the FeB phase with no Co content is indicated by the reflection from the (002) crystal face after removal of a material layer of 40 μm in thickness, too. What can be observed in addition is the variation of the crystal texture of the MeB phase from preferred orientation (111) in parallel with the specimen's surface for (Co, Fe)B with respect to a preferred orientation (002) for FeB. The X-ray diffraction diagram existing at 20 μm already includes the (002) reflection of the Fe_2B phase on the grounds of the tooth-like structure of the boride layer. With increasing depth, more reflections due to Fe_2B are established stepwise. The non-existance of reflections with (hk0) indication endorses the statement that Fe_2B has a (002) texture (MATUSCHKA). To demonstrate the impact of the Co content on the lattice constants, calculations were made to identify constants from reflections (Tab. 2). A significant variation of lattice constants is discernible for MeB, on the other hand, the lattice parameters calculated for Fe_2B are nearly constant.

Table 2

Lattice parameters of the (Co, Fe)B and Fe_2B phases calculated from X-ray diffraction diagrams of a boronized specimen (initial Co layer 20 μm) at different positions obtained by surface grinding

Phase	(Fe, Co)B			Fe_2B	
	lattice parameters				
	a	b	c	a	b
position/μm	10^{-10} m			10^{-10} m	
0	3.985	5.323	3.023		
20	3.979	5.332	3.029		4.25
40	4.021	5.423	2.983	5.11	4.26
50	4.025	5.427	2.980	5.12	4.25
60	4.071	5.527	2.957	5.12	4.29
70	4.10	5.49	2.96	5.13	4.26
80	4.08	5.52	2.95	5.13	4.25

When a specimen with the borides in the sequence (Co, Fe)B, FeB, Fe_2B (initial Co 10 μm, boronized by 1.6 vol.% BCl_3 in H_2 for one hour at 1223 K) is subjected to annealing in pure H_2 for one hour at 1223 K, thus, reflections of high relative intensity which may be attributed to the $(Co, Fe)_2B$ phase appear in the X-ray diffraction diagram. The relative intensity of reflections due to (Co, Fe)B and FeB, respectively, is significantly decreased. The conclusion from which fact could be that the major portion of (Co, Fe)B has already turned into $(Co, Fe)_2B$ after one-hour annealing, because Co_2B and Fe_2B exhibit the same crystal structure (I4/mcm (WYCKOFF)). The comparison of intensities further permits to presume that the (002) crystal texture which is generally known to exist for Me_2B (MATUSCHKA; POLI et al.; BOCHMANN et al. 1982, 1984, 1988) is reduced to the statistical distribution of crystallite orientation by the annealing effect.

4. Conclusions

All experimental data found indicate that the sequence of the phases in question is (Co, Fe)B, FeB, Fe_2B, substrate for the boronized specimens electroplated with cobalt. The lattice constants *a*, *b*, *c* of the (Co, Fe)B phase vary with decreasing cobalt content until FeB values are ultimately reached. (Co, Fe)B exhibits a preferred (111) orientation in parallel with the surface of the specimen whereas FeB displays a preferred (002) orientation. The (Co, Fe)B phase may be transformed into $(Co, Fe)_2B$ by annealing.

Electroplated cobalt coatings about 10 μm in thickness ensure that the substrate is sufficiently protected against any attacks made by the boronizing substances under the prevailing boronizing conditions and cause the surface of the boride layer to glitter in a dull shade with a uniform surface morphology brought about by relatively small particles. There boride layers are likely to present promising properties when in industrial use.

The authors thank the fellows of the structure laboratory, Sektion Physik/Elektronische Bauelemente, TU Karl-Marx-Stadt, who supported the preparation of SEM pictures, Dr. Kuechler and Mrs E. Rahm who carried out the EPXMA investigations, and Mrs I. Rothe who also involved in the said preparations.

References

BOCHMANN, G., SPÖRL, Th., RITZEL, B., WAGNER, W.: Wiss. Z. TH Karl-Marx-Stadt **24**, 682 (1982)
BOCHMANN, G., KLAUS, E., FRITSCHE, G., WAGNER, W.: Wiss. Z. TH Karl-Marx-Stadt **26**, 735 (1984)
BOCHMANN, G., RITZEL, B., WAGNER, W., MARX, G.: Cryst. Res. Technol. **20**, 1085 (1985)
BOCHMANN, G., KLAUS, E., FRITSCHE, G., WAGNER, W.: Wiss. Z. TU Karl-Marx-Stadt (in preparation) 1988
CARLSSON, J.-O.: Thin Solid Films **130**, 261 (1985)
DETTNER, H. W., ELZE, J.: Handbuch d. Galvanotechnik, Munich 1966, vol. 2
FICHTL, W.: Österreichische Ing.-Zeitschr. **22**, 426 (1979)
HEYDECKER, J., ZÖLLNER, H.: Z. wirtschaftliche Fertigung **75**, 391 (1980)
LEUSCHEN, K.: Ein Beitrag zur Feststoffborierung von Kobalt, Gußlegierungen auf Kobaltbasis und Sinterhartmetallen Dissertation Aachen 1977
MATUSCHKA, A., GRAF v.: Borieren, Munich 1977
MIRKIN, L. I.: Handbuch der Röntgenstrukturanalyse von Polykristallen, Moscow 1961
POLI, G., LANZONI, E., SAMBOGNA, G., POLOMBARINI, G.: Ceramica (Florence) **34**, 1 (1981)
RIEHLE, M.: Die Technik **32**, 682 (1977), **33**, 22 (1978)
WYCKOFF, R. W. G.: Crystal Structures, New York, London, Sidney 1965, vol. 1

(Received December 10, 1986)

Authors' address:

Dr. G. BOCHMANN, Chem.-Ing. E. KLAUS, Dipl.-Ing. G. FRITSCHE, Doz. Dr. W. WAGNER
Technische Universität Karl-Marx-Stadt, Sektion Chemie und Werkstofftechnik
PSF 964
Karl-Marx-Stadt
DDR-9010

Cryst. Res. Technol.	22	1987	7	969–976

K. D. Glinchuk, V. I. Guroshev, A. V. Prokhorovich

Institute of Semiconductors, Academy of Sciences of the Ukrainian SSR, Kiev

Irradiation Plus Annealing-induced 1.20 eV Emission Band in p-Type GaAs[1)]

The effect of 2.2 MeV electron irradiation and subsequent annealings on the luminescence of tellurium-compensated p-type GaAs crystals is studied and analyzed. A comparatively strong emission band peaked at $h\nu_m$ near 1.20 eV (induced by radiative electron recombination in $Te_{As}V_{Ga}$ pairs) appears in irradiated annealed (at $T \gtrsim 250$ °C) compensated p-type GaAs. The data obtained testify about the following: a) electron irradiation of GaAs creates stable (at $T \lesssim 200$ °C) defects not only in the arsenic sublattice of GaAs, but in its gallium sublattice too; b) radiation plus annealing (r.p.a.)-induced $Te_{As}V_{Ga}$ pairs are characterized by a rather high (compared to that for grown-in $Te_{As}V_{Ga}$ pairs) probability of electron radiative transitions in them. The electrical characterization of the r.p.a.-induced 1.20 eV radiative centres ($Te_{As}V_{Ga}$ pairs) is given.

Изучено и проанализировано влияние облучения электронами ($E = 2.2$ MeV) и последующих отжигов на люминесценцию компенсированных атомами теллура кристаллов p-GaAs. В облученном отожженном (при $T \gtrsim 250$ °C) компенсированном p-GaAs наблюдалась сравнительно интенсивная (индуцированная излучательной рекомбинацией электронов в парах $Te_{As}V_{Ga}$) полоса люминесценции с максимумом излучения $h\nu_m$ вблизи 1,20 eV. Полученные данные свидетельствуют о следующем: а) электронное облучение GaAs создает стабильные (при $T \lesssim 200$ °C) дефекты не только в мышьяковой, но и галлиевой подрешетке арсенида галлия; б) о достаточно высокой (превышающей соответствующую для ,,ростовых,, пар $Te_{As}V_{Ga}$) вероятности излучательных электронных переходов и индуцированных радиационо-термическим воздействием (РТВ) парах $Te_{As}V_{Ga}$. Приведены электрические характеристики созданных РТВ 1,20 eV излучающих центров (пар $Te_{As}V_{Ga}$).

1. Introduction

A study of radiation-induced emission bands in GaAs is quite promising to obtain an important information about irradiation-generated defects in gallium arsenide — the production, stability and characteristics of defects, on their origin, possible interaction with other defects, the sublattice they are mainly generated in and so on (Pons, Bourgoin; Glinchuk et al. 1986). The present paper is aimed at understanding of radiation-induced defects in GaAs using the luminescence technique as a method for their investigation. We will show that annealing at $T \gtrsim 250$ °C of electron-irradiated Te-compensated p-type GaAs results in an appearence of a rather strong in intensity emission band peaked at $h\nu_m$ near 1.20 eV. This will be taken as a proof that $Te_{As}V_{Ga}$ pairs are effectively formed in irradiated annealed GaAs crystals; the last testifies

[1)] Paper presented at the 2nd Conference on Physics and Technology of GaAs and other III–V semiconductors (Budapest/Hungary, September 8–11, 1986).

about generation by electron irradiation of stable at $T \lessapprox 200$ °C gallium vacancies[2]. Electrical characterization of radiation plus annealing (r.p.a.)-induced $Te_{As}V_{Ga}$ pairs will be given.

2. Experimental

Tellurium-compensated (and non-compensated) melt-grown copper (or zinc)-doped p-type GaAs crystals were used as a starting material. The donor (tellurium atoms) concentration $N_d \simeq (1 \text{ to } 4) \cdot 10^{17}$ cm^{-3}, the acceptor (copper or zinc atoms) concentration $N_a \simeq (2$ to $5) \cdot 10^{17}$ cm^{-3}; $N_d \lessapprox 1 \cdot 10^{16}$ cm^{-3} for uncompensated crystals. The free hole concentration p_0 was rather high (it was fixed by the concentration of uncompensated acceptors) at a room temperature — $p_0(300\text{ K}) \simeq N_a - N_d \simeq 1 \cdot 10^{17}$ cm^{-3} in zinc and copper-doped crystals. It was noticably less (due to deionization of uncompensated acceptors) at a liquid nitrogen temperature — $p_0(77\text{ K}) \simeq 1 \cdot 10^{16}$ $(2 \cdot 10^{10})$ cm^{-3} in zinc (copper)-doped crystals, accordingly. The initial (before irradiation) minority carrier (free electron) lifetime $\tau_n(0) \simeq$ $\simeq 10^{-10}$ s at 77 to 300 K.

Irradiation of GaAs by fast electrons of 2.2 MeV energy was made at 20 °C, the integrated (total) incident electron fluence $\Phi \approx 1 \cdot 10^{16}$ cm^{-2}. The irradiation-induced defect concentration $n_d \approx 5 \cdot 10^{16}$ cm^{-3}. Isochronal annealings of 15 min durations of irradiated (and control unirradiated) crystals were performed at several temperatures T^* between 100 and 550 °C. Electron irradiation and subsequent annealings did not change markedly the p_0-values and altered considerably the τ_n-values.

GaAs luminescence was excited by a strongly absorbed radiation (the excess carriers were concentrated at a depth $l \approx 0.3$ μm much less than the crystal thickness d) of a small-power He–Ne laser of intensity $L \lessapprox 1 \cdot 10^{19}$ quanta/cm^2 · s. Excitation did not change the excess free electron τ_n and hole τ_p ($\tau_n \ll \tau_p$) lifetimes ($\tau_{n,p} \neq f(L)$). The laser created maximum (near the GaAs surface $x = 0$) excess electron concentration $\delta n^* \approx L\tau_n/l \approx$ $\approx 3 \cdot 10^{13}$ cm^{-3} and the excess hole concentration $\delta p^* \approx L\tau_p/l \approx 5 \cdot 10^{14}$ cm$^{-3} \gg \delta n^*$ ($\delta p^* \gg p_0$ and $\delta p^* \ll p_0$ in copper- and zinc-doped crystals, accordingly).

In all crystals investigated (unirradiated, as irradiated) the luminescence killer (nonradiative) s-centres dominated in the excess carrier recombination (i.e. the free electron lifetime τ_n was not fixed by the radiative $Te_{As}V_{Ga}$ pairs) as the following confirms (see below): (a) the total luminescence intensity was much less than L; (b) no decrease in the free electron lifetime τ_n as the $Te_{As}V_{Ga}$ pair concentration N increased was observed ($\tau_n \neq f(N)$). The intensity I of the 1.20 eV emission band was measured by the calibrated PbS photodetector. Complex (linear, superlinear) emission intensities I vs exciting light intensities L dependences were observed in compensated p-type GaAs crystals[3]; so the $I(\Phi, T^*)$ dependences were measured in a region of a linear I vs L dependence. R.p.a.-induced variations in the concentration N of $Te_{As}V_{Ga}$ pairs were found from the analysis of corresponding variations in the intensities I and electron lifetimes τ_n (obviously, $I \sim$ $\sim N\tau_n$, see the Appendix, Eq. (A2)).

The electrical characterization of the r.p.a.-induced $Te_{As}V_{Ga}$ pairs (ε_d, ε_a, and α-values) was found from the analysis of temperature and excitation variations in the intensity of the 1.20 eV emission band (see the Appendix for details).

---

[2]) In GaAs a rather wide emission band peaked at $h\nu_m$ near 1.20 eV occurs due to radiative electronic transitions (the transition probability α) in neutral $Te^0_{As}V^0_{Ga}$ donor-acceptor pairs (the level energy position created by tellurium donors near the conduction band is ε_d and by gallium vacancies near the valence band is ε_a) (GLINCHUK, PROKHOROVICH).

[3]) Experimental data and some theoretical comment on the I vs. L dependences in compensated p-type GaAs see in the Appendix.

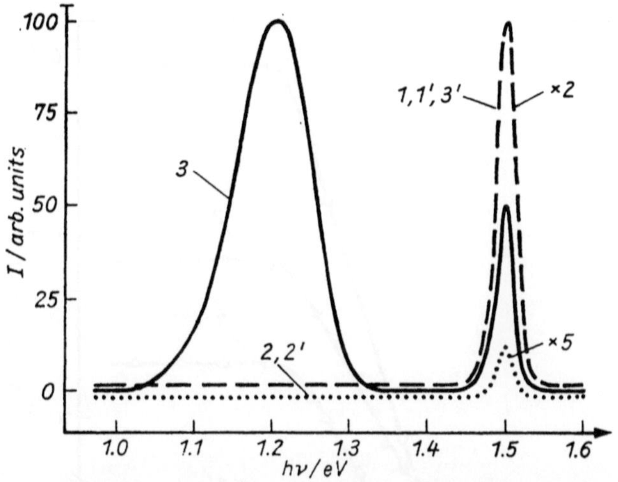

Fig. 1. Consecutive changes in the 77 K luminescence spectra in Te-compensated (1 to 3) and uncompensated (1' to 3') Zn-doped p-type GaAs crystals upon electron irradiation and subsequent annealings. (1, 1') before irradiation; (1, 1') after annealing at 200 to 400 °C for 15 min; (2, 2') after irradiation with 10^{16} electrons/cm²; (2, 2') after annealing at 200 °C for 15 min; (3, 3') after annealing at 400 °C for 15 min. $L = 10^{18}$ quanta/cm² · s

3. Results

Figures 1 and 2 show typical results of a study of luminescence in electron-irradiated and then 15 min isochronically annealed compensated (uncompensated) p-type GaAs crystals. The following main facts are clearly seen:

(i) As expected electron irradiation greatly decreased the intrinsic (peaked at $h\nu_m$ near 1.50 eV) emission intensity $I_{1.50}$ (the free electron lifetime τ_n) and subsequent annealing at $T \gtrsim 400$ °C practically restored the original (observed in unirradiated crystals) $I_{1.50}$ and τ_n-values (radiation-induced generation and annealing-induced annihilation of luminescence killer centres explain the above facts, GLINCHUK et al. 1986). The unexpected non-trivial fact is an appearence, as a result of annealing at $T \gtrsim 250$ °C of irradiated tellurium-compensated p-type crystals, of the comparatively wide (the halfwidth $W \approx 0.145$ eV at 77 K) emission band peaked at $h\nu_m$ near 1.20 eV (no effect is observed in electron-irradiated annealed uncompensated and control unirradiated annealed compensated p-type GaAs crystals).

(ii) An appearance of the above discussed 1.20 eV emission band is due to generation in electron-irradiated annealed at $T \gtrsim 250$ °C compensated p-type GaAs crystals of 1.20 eV radiative centres (their concentration N reaches the maximum value N_{max} at 400 °C; obviously, $N_{max} \lesssim n_d$).

(iii) The discussed r.p.a.-induced 1.20 eV emission band is rather strong in intensity — at maximum N-values nearly 10^{-3} of injected carriers recombine radiatively via 1.20 eV radiative centres ($I/L \approx 10^{-3}$).

4. Discussion

4.1. The stability of radiation-induced Ga vacancies in GaAs

The r.p.a.-induced appearance of 1.20 eV radiative centres in tellurium-compensated p-type GaAs crystals as the annealing temperature is raised above 200 °C is due to association (a capture) by tellurium atoms Te_{As} of radiation-induced mobile (at

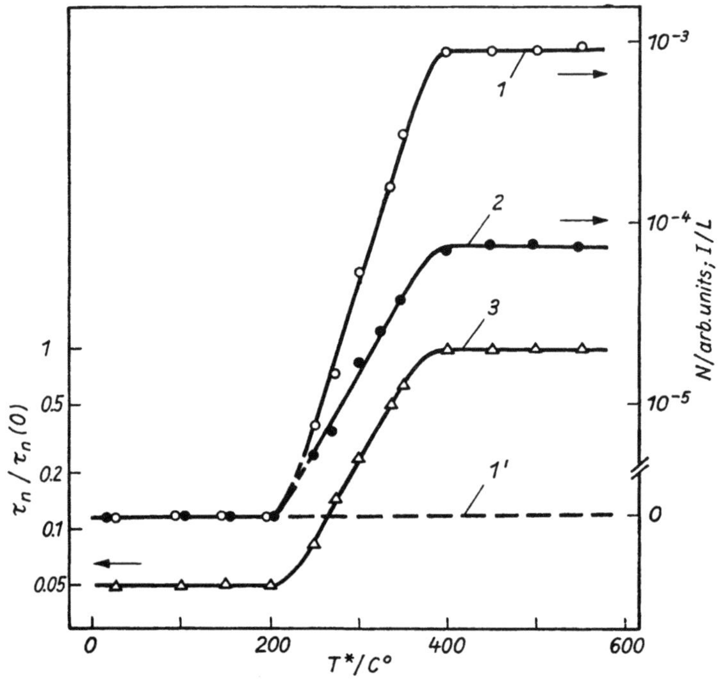

Fig. 2. Changes in the 77 K emission intensities I(1), concentrations of 1.20 eV radiative centres N (2) and 77 K free electron lifetimes τ_n (3) in Te-compensated Cu-doped p-type GaAs crystals upon electron irradiation with 10^{16} electrons/cm² followed by 15 min annealings at $T^* = 100$ to 550 °C. 1′ – $I(T^*)$ dependence for irradiated uncompensated Zn-doped p-type GaAs. $T^* = 20$ °C refers to unannealed crystals. $L = 10^{18}$ quanta/cm² · s

$T \gtrsim 200$ °C) gallium vacancies V_{Ga}. As a result $Te_{As}V_{Ga}$ pairs (they serve as 1.20 eV radiative centres, isolated gallium vacancies are not radiative centres themselves) appear according to the reaction:

$$Te_{As} + V_{Ga} \rightarrow Te_{As}V_{Ga} \, .$$

The aboved identification of the r.p.a.-induced 1.20 eV radiative centres in tellurium-compensated p-type gallium arsenide as $Te_{As}V_{Ga}$ pairs is confirmed by the following: (a) an appearance of the 1.20 eV emission band only in irradiated annealed Te-compensated crystals — no effect is observed in irradiated annealed uncompensated crystals (Fig. 1 and 2); (b) nearly the same $h\nu_m$-values for the r.p.a.-induced 1.20 eV radiative centres and $Te_{As}V_{Ga}$ pairs (Fig. 3).

The most important feature of the above pointed is the following. Previously (PONS, BOURGOIN) it was proposed that electron irradiation of GaAs creates stable (at $T \lesssim 200$ °C) defects only in the As sublattice of gallium arsenide (the radiation-induced defects in Ga sublattice of GaAs according to PONS and BOURGOIN are not stable ones-once created they anneal out at 4 K). The present results clearly show that stable (at $T \lesssim 200$ °C) radiation-induced defects in the Ga sublattice of gallium arsenide could also be produced by electron irradiation of GaAs (they clearly exist giving rise to generation at high annealing temperatures in irradiated crystals of gallium vacancy-related defects). In particular, as follows from the comparison of energy positions of the Ga vacancy level ε_a (see Sect. 4.2) and the Fermi level ε_F

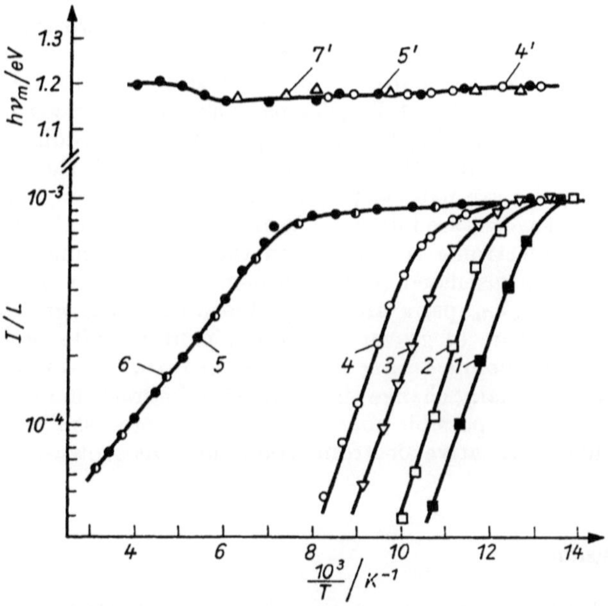

Fig. 3. Variations of intensities I (1 to 6) and emission peak energies $h\nu_m$ (4', 5') with reciprocal temperature for the r.p.a.-induced ($\Phi = 10^{16}$ electrons/cm², $T^* = 400$ °C) 1.20 eV emission band in Cu- (1 to 4, 4') and Zn-doped (5, 5', 6) Te-compensated p-type GaAs crystals. 7' — $h\nu_m(1/T)$ dependence for the extrinsic emission induced by grown-in $Te_{Ga}V_{As}$ pairs in Te-doped n-type GaAs. ε_d, eV: (1 to 6) $3 \cdot 10^{-3}$; ε_a, eV: (1 to 4) 0.12, (5, 6) 0.07: $L \cdot 10^{18}$ quanta/cm²s: (1) 0.4, (2, 5, 5') 1, (3) 4 and (4, 4', 6) 10

($\varepsilon_F \gtrless \varepsilon_a$) in p-type GaAs crystals at $T \lesssim 200$ °C stable (created by electron irradiation) neutral V_{Ga}^0 and negatively charged V_{Ga}^- gallium vacancies could really exist.

4.2. The donor and acceptor levels of the r.p.a.-induced $Te_{As}V_{Ga}$ pairs in GaAs

As the analysis shows (see the Appendix) the donor and acceptor levels (ionization energies) of the r.p.a.-induced $Te_{As}V_{Ga}$ pairs are: $\varepsilon_d \approx 3 \cdot 10^{-3}$ eV and $\varepsilon_a \approx 0.07$ to 0.12 eV.

As follows from the comparison of $h\nu_m$ and $E_g - \varepsilon_d - \varepsilon_a$ values ($h\nu_m \approxeq 1.2$ eV $< E_g - \varepsilon_d - \varepsilon_a \approx 1.4$ eV, here E_g is the band gap) and also W- and kT values ($W \gg kT$) the radiative electronic transition $\varepsilon_d \to \varepsilon_a$ in the r.p.a.-induced $Te_{As}^0V_{Ga}^0$ pairs is accompanied by the emission of phonons.

4.3. The probability of radiative electronic transitions in the r.p.a.-induced $Te_{As}V_{Ga}$ pairs

A comparatively high intensity of the r.p.a.-induced 1.20 eV emission band testifies about a rather high probability of radiative electronic transitions α in the r.p.a.-created $Te_{As}V_{Ga}$ pairs[4]). Really, $\alpha \approx 10^8$ s^{-1} for the r.p.a.-induced $Te_{As}V_{Ga}$ pairs as

[4]) The 1.20 eV emission band intensity I is rather high (depends weakly on temperature) at low temperatures and decreases considerably (is thermally quenched) at high temperatures (Fig. 3). An increase in the light intensity L could shift the region of a weak dependence of I on T and the emission thermal quenching curves to higher temperatures (Fig. 3). Some comment on the $I(1/T)$ and $I(1/T) = f(L)$ dependences see in the Appendix.

the corresponding estimates show[5]). It is interesting to note here that the aboved pointed α-value is much greater than the corresponding one for grown-in (formed during the GaAs growth) $Te_{As}V_{Ga}$ pairs — for the last $\alpha \approx 5 \cdot 10^5\ s^{-1}$ (GLINCHUK, PROKHOROVICH), i.e. the r.p.a.-induced $Te_{As}V_{Ga}$ pairs are characterized by the higher (compared to that for grown-in $Te_{As}V_{Ga}$ pairs) probability of radiative electronic transitions in them. Most probably the stated is connected with the following. The r.p.a.-induced $Te_{As}V_{Ga}$ pairs are naturally isolated ones (no other grown-in defects are located near them as the last are immobile at annealing temperature used); so the probability of radiative electronic transitions in them is determined only by the internal properties of pairs discussed (is not influenced by other defects). On the contrary the grown-in $Te_{As}V_{Ga}$ pairs are associated with other grown-in charged defects (the complex structure of grown-in $Te_{As}V_{Ga}$ pairs is quite probable as the grown-in defects are highly mobile during the crystal growth, GLINCHUK et al. 1981). The associated defects essentially change the Coulombic interaction of electrons with the grown-in $Te_{As}V_{Ga}$ pairs; a possible consequence of the process discussed is a decrease in the probability of radiative electronic transitions via grown-in $Te_{As}V_{Ga}$ pairs (GLINCHUK et al. 1981).

5. Conclusion

The data presented about generation by electron irradiation in GaAs of stable defects in its Ga sublattice and about a rather high probability of electronic radiative transitions in the r.p.a.-induced $Te_{As}V_{Ga}$ pairs are important for understanding the complex problem of radiation-induced defect production in compound intermetallic semiconductors which is widely studied now (see, for example, Pons, Bourgoin).

Appendix

I vs *T* and *L* dependences

Below some comment on the temperature and excitation variations in the r.p.a.-induced 1.2 eV emission intensities for p-type GaAs will be given.

Let us find the intensity of the 1.20 eV emission occuring due to radiative recombination of excess carriers in $Te_{As}V_{Ga}$ pairs in p-type GaAs. Rather slow electronic transitions from donors to acceptors $\varepsilon_d \rightarrow \varepsilon_a$ and comparatively rapid electronic transitions from donors to the conduction band $\varepsilon_d \rightarrow E_c$ and from acceptors to the valence band $\varepsilon_a \rightarrow E_v$ take place in the aboved pairs (Fig. 4). So: a) the filling of paired donor levels by electrons is mainly determined by their exchange with the conduction band (the corresponding filling probability $\varphi_d \approx \delta n/(\delta n + n_1)$, where $n_1 = Q_c \exp(-\varepsilon_d/kT)$; b) the filling of paired acceptor levels by holes is mainly determined by their exchange with the valence band (the corresponding filling probability $1 - \varphi_a \approx (p_0 + \delta p)/(p_0 + \delta p + p_1)$, where $p_1 = Q_v \times \exp(-\varepsilon_a/kT)$)[6]). The discussed emission occurs due to radiative donor-to-acceptor electron transitions in neutral $Te^0_{As}V^0_{Ga}$ pairs; as a result, the last are transformed into $Te^+_{As}V^-_{Ga}$ pairs (Fig. 4). Therefore, the intensity of the emission caused by $Te_{As}V_{Ga}$ pairs obviously is determined by the concentration of neutral pairs $Te^0_{As}V^0_{Ga}$ δN^0 (undoubtedly, for the case discussed $\delta N^0 \approx \varphi_d(1 - \varphi_a)\ N$) and by the probability of radiative electroni transitions in the aboved pairs α. Let: (a) the $Te_{As}V_{Ga}$ pairs do not prevail in the excess carrier recombination ($I \ll L$ and $\tau_n \neq f(N)$), i.e. the linear ($\tau_n \neq f(L)$) volume non-radi-

[5]) The α-value was evaluated from the relation (A2) using the known I, $N_{max} \approx 0.5 n_d$, τ_n and ε_d values (see the Appendix).

[6]) Q_c and Q_v are effective densities of conduction band and valence band states ($Q_{c,o} \sim T^{3/2}$).

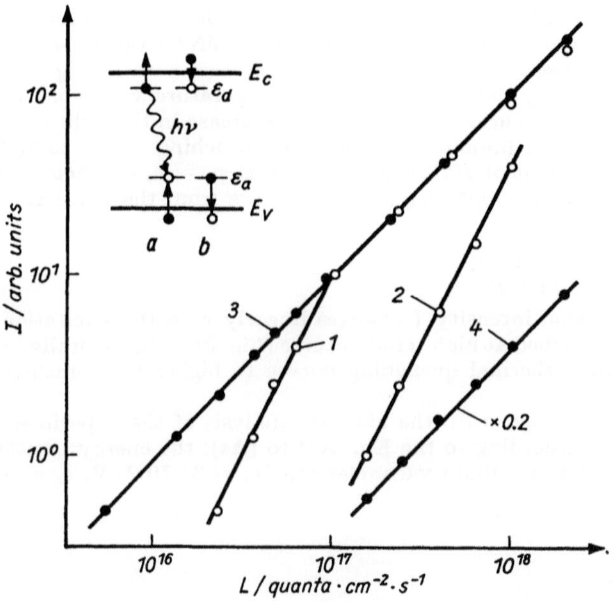

Fig. 4. Intensity of the r.p.a.-induced ($\Phi = 10^{16}$ electrons/cm², $T^* = 400$ °C) 1.20 eV emission band I vs exciting light intensity L for Cu- (1, 2) and Zn-doped (3, 4) Te-compensated p-type GaAs crystals. T, K: (1, 3) 77, (2) 100 and (4) 200. In an inset a scheme of electronic transitions via $Te^0_{As}V^0_{Ga}$ (a) and $Te^+_{Ga}V^-_{As}$ (b) pairs is shown

ative recombination of excess carriers via deep s-centres dominate[7]); (b) the extrinsic 1.20 eV emission is excited by the strongly absorbed light of intensity L, excess electron and hole concentrations decay exponentially with the distance x in the semi-infinite ($l \ll d$) GaAs crystal — $\delta n(x) \approx \delta n^* \exp(-x/l)$, $\delta p(x) \approx \delta p^* \exp(-x/l)$; (c) excitation creates rather low δn-values ($\delta n(x) \ll n_1$) and any δp-values (in this case $\varphi_d(x) \approx \delta n(x)/n_1 \ll 1$, $1 - \varphi_a(x) \lesssim 1$ and $\delta N^0(x) \ll N$). Then the 1.20 eV emission intensity I ($\ll L$) could be easily found by integration from zero to infinity of "local" 1.20 eV emission intensities $i(x) = \alpha \delta N^0(x) \approx \alpha \varphi_d(x) [1 - \varphi_a(x)] N$ over the crystal depth x:

$$I = \int_0^\infty i(x)\, dx \approx \alpha N \frac{L\tau_n}{n_1} \left[1 - \frac{p_1}{\delta p^*} \ln\left(1 + \frac{\delta p^*}{p_0 + p_1}\right)\right]. \tag{A1}$$

At low temperatures (in particular, at 77 K) $p_1 \ll p_0 + \delta p^*$, so, as follows from the Eq. (A1) the emission intensity:

$$I \approx \alpha N(L\tau_n/n_1) = \alpha N(L\tau_n/Q_c) \exp(\varepsilon_d/kT)\,. \tag{A2}$$

In this case the emission intensity is rather high (the paired gallium vacancies are mainly filled by holes, i.e. $1 - \varphi_a \approx 1$ and so, δN^0 is comparatively high) and linearly increases with the excitation intensity (Fig. 4); the emission intensity temperature variations are mainly determined by the corresponding variations in δN^0 due to φ_d-changes (Fig. 3).

At high temperatures $p_1 \gg p_0 + \delta p^*$, so, as follows from the Eq. (A1) the emission intensity:

$$I \approx \alpha N \frac{L\tau_n}{n_1} \cdot \frac{p_0 + 0.5\delta p^*}{p_1} = \alpha N \frac{L\tau_n}{Q_c} \cdot \frac{p_0 + 0.5\delta p^*}{Q_v} \exp\left(\frac{\varepsilon_d + \varepsilon_a}{kT}\right). \tag{A3}$$

---

[7]) Obviously, if the $Te_{As}V_{Ga}$ pairs prevail in the excess carrier recombination (the excess carriers recombine mainly via the above pairs, i.e. τ_n is determined by the last), then $I \approx L$ at any pair concentrations and temperatures.

In this case the emission intensity falls considerably (the paired gallium vacancies become weakly occupied by holes, i.e. $1 - \varphi_a \ll 1$, and so, δN^0 is rather small) and linearly (at $\delta p^* \ll p_0$) and quadratically (at $p_0 \ll \delta p^* \sim L$) increases with L (Fig. 4); the emission thermal quenching is mainly determined by the temperature variations in δN^0 due to correspouding φ_d and $1 - \varphi_a$ variations (Fig. 3). An increase in the light intensities L (i.e. in $p_0 + \delta p^*$-values) shifts the luminescence thermal quenching curves to higher temperatures as the "inflection" point of $I(T)$ dependence is at $p_1 \approx p_0 + 0.5\delta p^*$ (Fig. 3).

At low excitation intensities $\delta p^* \ll p_0$, so, as follows from the Eq. (A1) the emission intensity is:

$$I \approx \alpha N \frac{L\tau_n}{n_1} \cdot \frac{p_0}{p_0 + p_1}. \tag{A4}$$

In this case the emission intensity I changes linearly with the excitation intensity L (Fig. 4), is thermally quenched at high temperatures (Fig. 3), and, naturally, no excitation-induced shift of emission thermal quenching curves to higher temperatures is observed (Fig. 3).

As follows from the analogous (to the aboved) analysis of the experimental $I(1/T, L)$ data for $Te_{As}V_{Ga}$ pairs (according to the Eq. (A2) to (A4)) the energy positions of paired tellurium donors and paired gallium vacancies are: $\varepsilon_d \approx 3 \cdot 10^{-3}$ eV, $\varepsilon_a \approx 0.07-0.12$ eV (Fig. 3).

References

GLINCHUK, K. D., PROKHOROVICH, A. V.: phys. stat. sol. (a) **41**, 659 (1977)

GLINCHUK, K. D., LUKAT, K., PROKHOROVICH, A. V.: phys. stat. sol. (a) **64**, K139 (1981)

GLINCHUK, K. D., GUROSHEV, V. I., PROKHOROVICH, A. V.: Cryst. Res. Technol. **21**, 811 (1986)

PONS, D., BOURGOIN, G. C.: J. Phys. C **18**, 3839 (1985)

(Received December 2, 1986)

Authors' address:

Dr. K. D. GLINCHUK, V. I. GUROSHEV, A. V. PROKHOROVICH
Institute of Semiconductors, Academy of Sciences of the Ukrainian SSR
prospect Nauki 115
USSR, 252028 Kiev 28

Cryst. Res. Technol.	22	1987	7	977–980

P. VENUGOPAL REDDY, K. J. PRATAP, T. SESHAGIRI RAO

Department of Physics, University College of Science,
Osmania University, Hyderabad, India

Electrical Conductivity of Some Mixed Ferrites at Curie Point

The behaviour of electrical conductivity of four ferrite semiconductor systems viz., Li–Ni, Mn–Mg, Co–Zn and Ni–Zn having different compositions has been studied in the vicinity of magnetic transformation temperature. Contrary to the predictions of an empirical formula (PARKER's) all the ferrites under investigation have exhibited marked and clear transitions at the Curie Point.

1. Introduction

Ferrites are semiconductors by nature. It is established that the electrical phenomenon in ferrites especially in the neighbourhood of magnetic transformation temperature is very complicated one and is not fully understood. For example, in some cases it is observed that the conductivity versus temperature curves exhibit sharp, clear and marked transitions at the Curie Point, while in some other cases the transitions areϱ either very small or not visible at all. Thus it would be extremely interesting, important and essential to study the electrical conductivity of ferrite semiconductors in the neighbourhood of magnetic transition temperature. As such, a study of electrical conductivity at Curie point of mixed Li–Ni, Mn–Mg, Co–Zn and Ni–Zn ferrites having different compositions has been undertaken and the results are presented in the paper.

2. Experimental

The polycrystalline ferrites of the four mixed ferrite systems, viz., Li–Ni, Mn–Mg, Co–Zn, and Ni–Zn were prepared by the conventional double sintering process. The details of the method of preparation have been reported earlier from this laboratory by REDDY et al. (1977, 1980), MURTHY et al. and REVATHI RAO. The DC electrical conductivity for all the samples has been determined over a temperature range 300 to 900 K by the well-known two-probe method (REDDY et al. 1981). Plots of log σT versus $10^3/T$ for all the specimens have also been reported earlier from this laboratory by REDDY et al. (1981 a, b) and SATYANARAYANA et al. (1982, 1983).

Thermoelectric power studies have also been undertaken with a view to know wehther the ferrites under study are n-type or p-type semiconductors.

3. Results and discussion

The compositional formulae of all the ferrites along with the resistivity values at Curie point i.e., ϱ_θ and Curie temperature are given in Tables 1–4. It can be seen

Table 1
Electrical resistivity data of mixed Li–Ni Ferrites

Specimen No	composition	curie point θ/K	resistivity at Curie point ϱ_θ/Ohms · cm
LN-I	$Li_{0.5}Fe_{2.5}O_4$	870	9.90×10^3
LN-II	$Lo_{0.47}Ni_{0.06}Fe_{2.47}O_4$	848	9.35×10^1
LN-III	$Li_{0.35}Ni_{0.3}Fe_{2.35}O_4$	830	7.30×10^4
LN-IV	$Li_{0.2}Ni_{0.6}Fe_{2.2}O_4$	500	1.00×10^5
LN-V	$Li_{0.05}Ni_{0.9}Fe_{2.05}O_4$	880	1.71×10^2
LN-VI	$NiFe_2O_4$	863	1.32×10^3

Table 2
Electrical resistivity values at Curie Point of mixed Mn–Mg ferrites

Specimen No	composition	Curie point θ/K	resistivity at Curie point ϱ_θ/Ohms · cm
MM-I	$MnFe_2O_4$	573	2.60×10^4
MM-II	$Mn_{0.75}Mg_{0.25}Fe_2O_4$	623	1.75×10^4
MM-III	$Mn_{0.5}Mg_{0.5}Fe_2O_4$	543	4.08×10^3
MM-IV	$Mn_{0.25}Mg_{0.75}Fe_2O_4$	643	1.64×10^4
MM-V	$Mn_{0.1}Mg_{0.9}Fe_2O_4$	623	4.81×10^2
MM-VI	$MgFe_2O_4$	710	8.62×10^4

Table 3
Electrical resistivity values at Curie point of mixed Ni–Zn ferrites

Specimen No	composition	Curie point θ/K	resistivity at Curie point ϱ/Ohms · cm
NZ-I	$Ni_{0.9}Zn_{0.1}Fe_2O_2$	798	8.55×10^2
NZ-II	$Ni_{0.7}Zn_{0.3}Fe_2O_4$	688	3.45×10^4
NZ-III	$Ni_{0.5}Zn_{0.5}Fe_2O_4$	573	3.13×10^4
NZ-IV	$Ni_{0.3}Zn_{0.7}Fe_2O_4$	381	2.06×10^5

Table 4
Electrical resistivity values at Curie point of mixed Co–Zn ferrites

Specimen No	composition	Curie point θ/K	resistivity at Curie point ϱ_θ/Ohms · cm
CZ-I	$CoFe_2O_4$	783	3.17×10^4
CZ-II	$Co_{0.8}Zn_{0.2}Fe_2O_4$	647	7.35×10^2
CZ-III	$Co_{0.5}Zn_{0.5}Fe_2O_4$	543	2.48×10^3
CZ-IV	$Co_{0.3}Zn_{0.7}Fe_2O_4$	361	1.09×10^6

from the tables that the values of resistivity at Curie point i.e. $\varrho_\theta > 10$ Ohm · cm for all the specimens. It was also observed earlier by REDDY et al. (1981 a, b), and SATYANARAYANA et al. (1982, 1983) that all the specimens show clear and marked transitions at the magnetic transformation temperature where a ferrimagnetic material becomes a paramagnetic one.

It was pointed out by PARKER (1958, 1975) that if an anomaly in the conductivity versus temperature curve occurs near the Curie temperature, the result of the spin order is generally to lower the resistivity. The author further pointed out that the criterion for an influence of spontaneous magnetization on electrical conduction is the value of the resistivity at the magnetic transformation temperature i.e. ϱ_θ. The author classified the magnetic materials into three categories. An anomaly is always observed if $\varrho_\theta \leqq 0.5$ Ohm · cm (Category III) and if $\varrho_\theta \geqq 10$ Ohm · cm (Category I). the electrical conductivity remains unaffected and no transition is possible at Curie point. Between these two values of conductivity (Category II) apparently either type is possible. The author experimentally arrived at the result which states that, as the values of resistivity at Curie point (ϱ_θ) for nickel-, copper-, manganese-, and magnesium ferrites are greater than 10 Ohm · cm, the magnetic transitions were not observed. However, REDDY et al. (1983, 1984) earlier reported that eventhough $\varrho_\theta = 1.32 \times \times 10^3$ Ohm · cm in the case of nickel ferrite and $\varrho_\theta = 1 \times 10^3$ Ohm · cm in the case of magnesium ferrite it is possible to obtain predominant transitions at the Curie point.

It can be seen from the tables that the values of resistivity at Curie point (ϱ_θ) for all the samples under study are greater than 10 Ohm · cm. As such, according to PARKER's theory (PARKER 1958, 1975) none of the ferrites under study should exhibit any kind of transition at the Curie point. Contrary to this theory it is observed experimentally beyond any doubt that all the ferrites exhibited clear and marked transitions. This clearly indicates that the criterion for observing transitions at the Curie point may not be the value of resistivity at the magnetic transformation temperature as suggested by PARKER. Further studies are in progress to arrive at a suitable explanation for the phenomenon.

One of the authors (PVR) is grateful to Dr. R. SATYANARAYANA for useful discussions.

References

MURTHY RAMANA, S., REVATHI, B., RAO, T. S.: J. Less-Comm. Met. 57, 742 (1977)
PARKER, R.: Phil. Mag., 3, 853 (1958)
PARKER, R.: Electrical properties of magnetic oxides (D. J. CRAIK, Ed.), London 1975
REDDY VENUGOPAL, P., RAMANA, Y. V., RAO, T. S.: Current Science 46, 742 (1977)
REDDY VENUGOPAL, P., RAO, T. S.: J. Less-Comm. Met. 75, 255 (1980)
REDDY VENUGOPAL, P., RAO, S. M. D., RAO, T. S.: J. Less-Comm. Met. 79, 191 (1981) a
REDDY VENUGOPAL, P., RAO, S. M. D., RAO, T. S.: J. Less-Common. Met. 79, 1 (1981) b
REDDY VENUGOPAL, P., SATYANARAYANA, R., RAO, T. S.: phys. stat. sol. (a) 78, K109 (1983)
REDDY VENUGOPAL, P., RAO, T. S.: J. Mat. Sci. Lett. 3, 847 (1984)
REVATHI, B., RAO, T. S.: J. Less-Comm. Met. 57, 29 (1978)

SATYANARAYANA, R., MURTHY RAMANA, S., RAO, T. S.: J. Less-Comm. Met. 86, 115 (1982)
SATYANARAYANA, R., MURTHY RAMANA, S., RAO, T. S.: J. Less-Comm. Met. 90, 243 (1983)

(Received December 11, 1986)

Author's address:

Dr. P. VENUGOPAL REDDY*), Dr. T. SESHAGIRI RAO, K. J. PRATAP
Department of Physics, Osmania University,
Hyderabad-500007, A.P., India

*) *Present address*:

University College of Technology,
Osmania University,
Hyderabad-500007, India

Cryst. Res. Technol.	22	1987	7	981–986

J. W. Tomm, B. Sumpf, Karin Herrmann, A. Szczerbakow

Humboldt-Universität zu Berlin, Sektion Physik
Instytut Fizyki, PAN., Warszawa

Some Aspects of the Technology of Lead Salt Diode Lasers Used in Gas Monitoring Systems

A rather simple method of lead salt diode laser fabrication is described. The optimization of the diffusion process forming the pn-junction is especially discussed with regards to the various carrier concentrations of the as-grown crystals. Liquid nitrogen cooled lead salt diode lasers of the described type meet all demands of a laser based air pollution monitoring device.

Es wird eine verhältnismäßig einfache Methode der Bleisalzdiodenlaserherstellung erläutert. Insbesondere die Frage der Optimierung des Diffusionsprozesses, der den pn-Übergang erzeugt, in Abhängigkeit von der Ausgangsladungsträgerkonzentration der Kristalle wird diskutiert. Halbleiterinjektionslaser des beschriebenen Typs entsprechen voll allen Anforderungen, die an ein mit flüssigem Stickstoff gekühltes, mit Bleisalzlasern arbeitendes Schadstoffmeßgerät gestellt werden.

The application of lead salt lasers for spectroscopical and analytical research is well-known (see e.g. Preier; Linden; Kosichkin, Nadezhdinskij).

Reports on maximum operating temperatures near room-temperature (Partin) let us hope that in the future uncooled lead-salt quantum well lasers will be available for spectroscopic purposes. But we have also to note that those devices are still too complicated and expensive for the next few years.

A number of important applications is connected with demands to the device parameters which are realized using more simple laser structures. In this paper we describe a rather simple technology for $Pb_{1-x}Sn_xSe$- and $PbS_{1-x}Se_x$ homojunction lasers. Devices of this type are used for air pollution detection (CO, CO_2, SO_2, NO_X, ...) in long path measurements.

A description of the experimental equipment is given by Herrmann et al. and the computer-coupling is given in the paper by Sumpf and Göring.

The application of pulsed lead salt lasers for air pollution detection in free atmosphere requires the following features of the device:

— maximum operating temperature of about 100 K (liquid nitrogen cooling possible)
— threshold current significantly smaller than 5 A.
— spectral tuning rate within a laser mode better than 0.1 $cm^{-1}\,\mu s^{-1}$
— time-constant of the detector-amplifier system in the order of 100 ns
— signal-to-noise-ratio of the detector-amplifier system better than 100:1 (without boxcar integrator processing).

Practice showed that broad area laser devices which fulfil these demands in pulsed operation, have maximum cw-working temperatures less than 60 K.

Stripe geometry devices should not be used for pulsed applications because of their larger tuning rate. Highly reliable $Pb_{1-x}Sn_xSe$- and $PbS_{1-x}Se_x$ homojunction lasers were produced with the technology described in this paper. Homogeneous crystals for this purpose were grown using a sublimation method according to the rules given by SZCZERBAKOW. Some modifications were applied. Lead, tin, sulphur, and selenium with the purity of 5 N were used as the starting materials for crystal preparation. After stoichiometrical mixing the single elements, the binary components were synthesized in a quartz tube using zone floating apparatus. Further homogenization was completed by heating the quartz tube for 200 hours at 1070 K. Mixtures of binary components used as the source material in crystal growth processes were prepared according to the desired x-ratio which is determined by the absorption line position of the given gas.

Crystals grown with the method described usually have carrier concentrations of p_0 or n_0 in the range from 10^{17} to 10^{19} cm^{-3} and mobilities μ from some 10^3 to some 10^4 cm^2(Vs)$^{-1}$ at 80 K. For the following preparation of diode lasers only n-type crystals with mobilities better than $5 \cdot 10^3$ cm^2(Vs)$^{-1}$ were used.

p-type samples easily may be definitively converted into n-type by a secondary crystal growing with an excess of lead. In our experiments we found, that the number of electrons delivered by one Pb atom is in the order of two. Taking into account the results of PALMETSHOFER we assume that Pb atoms are built into lead vacancies. The mobility of the converted material was always smaller than that of the primary crystal. Therefore we assume that the mobility is limited by foreign impurities. Using n-type crystals with the characteristical parameters mentioned above the diffusion was realized in saturated Se vapour in a quartz tube. The equipment used for the diffusion allows to control independently the temperature of the crystal (T_2) and the Se source (T_1) (that means the vapour pressure). This two zone method of crystal growth was described earlier by BARTHEL et al.

To select the best diffusion conditions a certain number of principles should be fulfilled.

1. The optical active layer of the laser should have a small majority carrier concentration to reduce the losses due to absorption of free carriers and to Auger recombination. The experimental proof of this thesis was given by SHINOHARA et al.
2. For pn-junctions formed by diffusion one should prefer diffusion techniques providing the largest concentration gradients in the junction range.
3. Se overlayers on the crystal surface should be reduced to avoid problems connected with the properties of the contact layers.

To understand the theoretical background of the diffusion process we applied the model of interdiffusion. Using the results of GULDI et al. concerning the diffusion coefficients in the case of PbSe we extend the theory to the whole $PbS_{1-x}Se_x$-system.

Fg. 1. $PbS_{0.68}Se_{0.32}$ single crystal ingot (1 mm unit)

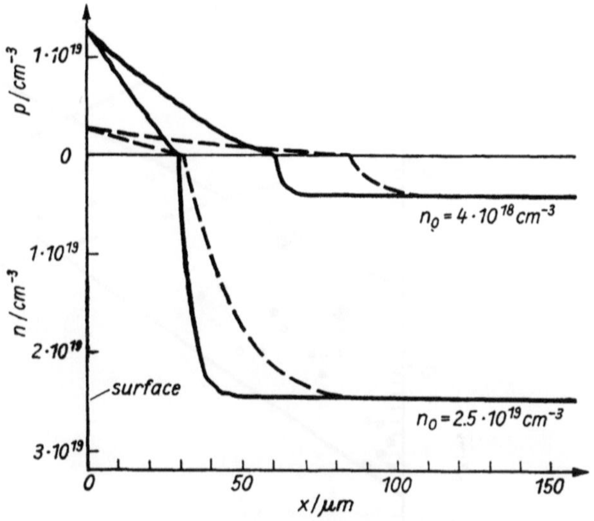

Fig. 2. Linear plots of the calculated diffusion profiles at the surface of the crystal. ——— Isothermal diffusion (i) ($T_1 = 770$ K, $T_2 = 770$ K, $t = 30$ min); – – – Two-zone diffusion (ii) ($T_1 = 600$ K, $T_2 = 770$ K, $t = 3$ h)

For calculations of the concentration profile we assume an average diffusion coefficient within the p- and n-region, resp. and an untiring source. The applicability of these theoretical results was shown by measurements of the local excited photoeffects which allowed to determine the depths of the p–n junction and the diffusion lengths. These results will be discussed elsewhere by KÜHNEMANN et al.

For a great number of experiments we varied the diffusion parameters (T_1, T_2 and the diffusion time t). Some of the calculated profiles for lasers are shown in Figure 2. The results for the junction depths are in good agreement with the experiment. The following two methods ensure the best laser parameters.

(i) Isothermal diffusion

($T_1 = 770$ K, $T_2 = 770$ K, diffusion time $t = 30$ min)
The experimental proof that this set of diffusion parameters gives lasers with low threshold current densities, if the substrates had low carrier concentrations, was given by SHINOHARA. We reproduced these results given in Figure 3 (full points). Own results are added (full circles). Compared with (ii) the concentration gradient near the junction is here larger and the value of p obtained by Se diffusion is higher. One can speculate that the insufficient results for crystals with a higher n_0-concentration are due to the fact that there is not any layer near the junction which has a sufficiently low concentration to become an optical active layer.

For low n_0 values the optical active layer seems to be situated in the n-region. From our measurements we can conclude, that the diffusion parameters mentioned above only can be used for crystals with low n_0. Furthermore Se overlayers have been observed on the crystal surface.

(ii) 2-zone diffusion

($T_1 = 600 \ldots 630$ K, $T_2 = 770$ K, diffusion time $t = 3$ h)
From the calculations plotted in Figure 2 follows, that for $T_1 = 600 \ldots 630$ K p is reduced compared with (i). For this set of diffusion parameters better lasers are obtained for crystals with high n_0 values (see Figure 3, open circles) and with lower energy gaps ($PbS_{1-x}Se_x$ ($x \lesssim 1$) and $Pb_{1-x}Sn_xSe$).

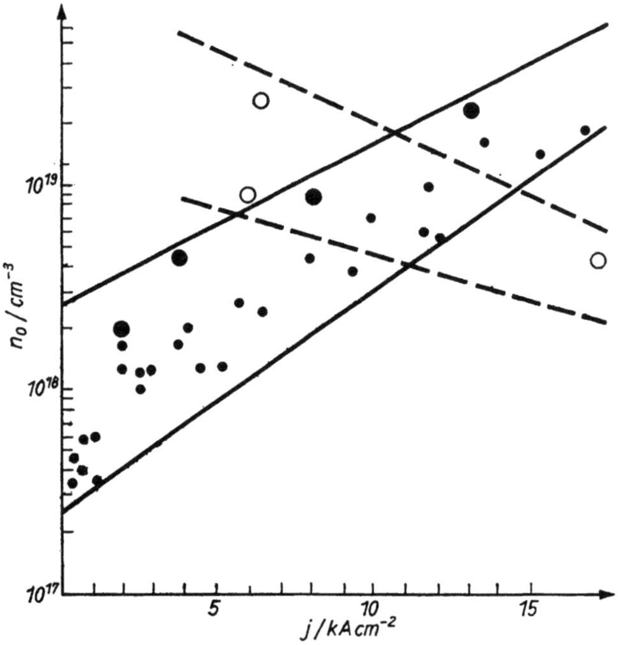

Fig. 3. Pulsed threshold current densities with respect to carrier densities n_0 of the substrates. ● SHINOHARA et al.; ⬤ $PbS_{0.68}Se_{0.32}$ ($T_1 = T_2 = 770$ K, $t = 30$ min) (i); ○ $PbS_{0.68}Se_{0.32}$ ($T_1 = 600$ K, $T_2 = 770$ K, $t = 3$ h) (ii)

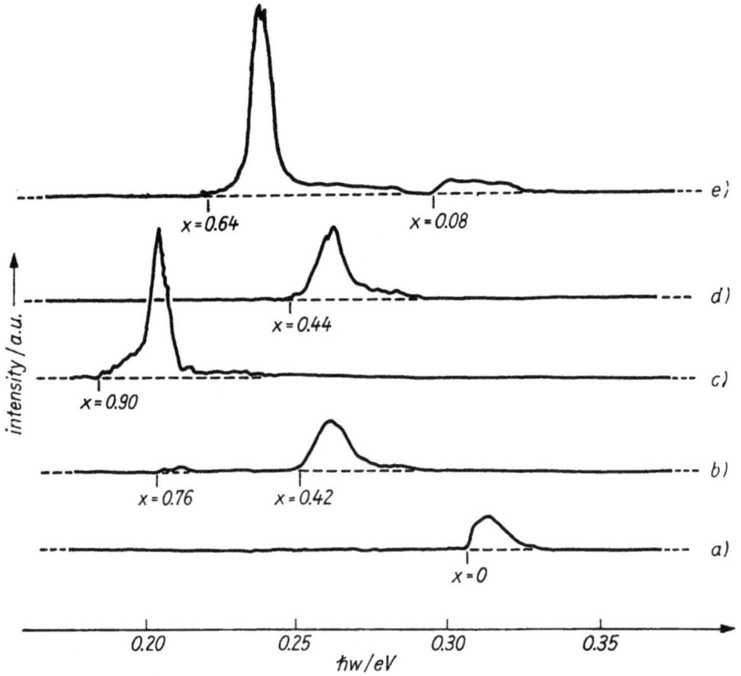

Fig. 4. Photoluminescence spectra at 80 K of a PbS single crystal before (a)) and after (b) ... (e)) the diffusion procedure. Using the well-known formula given by PREIER the x-value corresponding to the low energetical limit of the spectra (E_g) is given

According to (i) it seems that now the p-side of the junction becomes the optical active layer. But we have also to note, that the concentration gradient on both sides of the junction range becomes smaller. Se overlayers on the crystal surface are reduced by the two-zone diffusion technique but not excluded. The influence of Se overlayers on the crystals is shown in photoluminescence. Figure 4a shows a photoluminescence spectrum (PL) obtained from a pure PbS crystal. Figures 4 b ... e show PL spectra taken from the same crystal (various places on the surface) after a two-zone diffusion. From Figure 4 follows that not only Se overlayers occur on the crystal surface, but even macroscopic large surface regions (several tenths of microns) obtain the character of mixed crystals with a more or less defined gap value. This effect is limited to the surface of the crystals only, because in case the laser line position was in the spectral region expected from the x-value.

A number of experiments was carried out to find differences in the threshold current of lasers made by diffusion on as-grown and cleaved surfaces.

In the low temperature range ($T < 40$ K) lasers fabricated by diffusion on as-grown surfaces have lower threshold currents. For temperatures $T > 60$ K threshold current

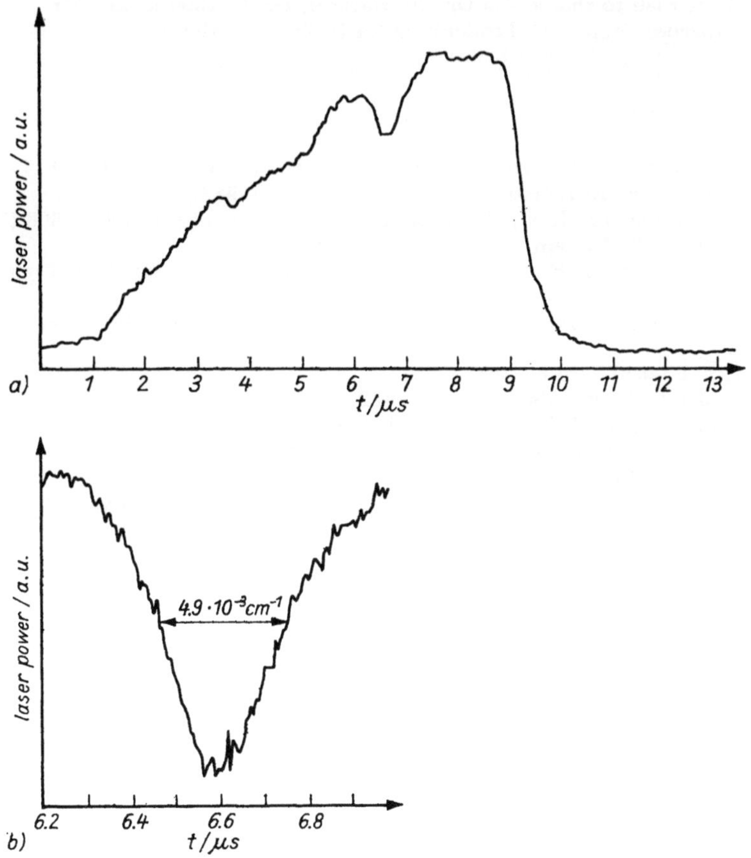

Fig. 5. Doppler broadened CO-line scanned with a $PbS_{0.68}Se_{0.32}$ diode laser in an air pollution test device. (parameters of laser operation: pulsed, $\tau = 10\ \mu s$, $f = 100$ Hz, current 3 A, $T = 80$ K). The laser was produced by the technique given in this paper: a) Pulsed laser signal; b) Enlargement of the line situated in a) between 6 and 7 μs

of lasers made by diffusion on cleaved surfaces were not higher than that of lasers made by diffusion as-grown surfaces.

One should note that this result allows to use significantly more surfaces from every crystal for diffusion and laser preparation, if low threshold currents are demanded only at the temperature of liquid nitrogen, because cleaving into several wafers parallel to the as-grown surface before diffusion becomes possible.

After the diffusion procedure a gold layer was deposited on the crystal surface (HERRMANN et al.). Cleaving and the formation of contacts was made using conventional techniques. A number of control measurements were made parallel to the technique mentioned above. Hall effect measurements delivered the values for n_0 and p_0 and the mobilities μ. PL measurements delivered relative values for the rate of radiative recombination, for the given x-value and overlayers, resp. Tests with a laser probe gave information about junction depths. In Figure 5 a medium resolution spectrum of a Doppler-broadened CO line is plotted and the operation conditions of the laser are given. Highly reliable lasers of that type completely meet all demands of air pollution detection.

The authors wish to thank Mrs Dr. Ch. Barthel, Dr. U. Barthel and Dr. A. E. Yunovich for helpful discussion, and G. Lindenberg for technical assistance.

References

BARTHEL, C., BARTHEL, U., TOMM, J. W., SICHE, D., KARCZEWSKI, G.: 8. Tagung Physik und Elektronik, Kurzfassungen der Vorträge 23 (1983)
GULDI, R. L., WALPOLE, J. N., REDIKER, R. H.: J. Appl. Physics: **44**, 4896 (1973)
HERRMANN et al.: Z. Meteorologie (to be printed)
HERRMANN, KA., SUMPF, B., BÖHME, D., HANNEMANN, M.: Cryst. Res. Technol. **18**, 1083 (1983)
KOSICHKIN, YU. V., NADEZHDINSKIJ, A. I.: Izv. Akad. Nauk SSSR, Ser. Fiz. **47**, 2037 (1983)
KÜHNEMANN, F. et al.: (to be published)
LINDEN, K. J.: J. Electr. Mat. **11**, 575 (1982)
PALMETSHOFER, L.: Appl. Physics **34**, 134 (1984)
PARTIN, D. L.: Superlattices and Microstructures **1**, 131 (1985)
PREIER, H.: Appl. Physics **20**, 189 (1979)
SHINOHARA, K., ITO, M., FUKUDA, H., UEDA, R.: Jap. J. Appl. Physics **20**, 4, 779 (1981)
SUMPF, B., GÖRING, D.: Exp. Math. and wiss. Gerätebau 86/1 (1986)
SZCZERBAKOW, A.: J. Cryst. Growth (in print)

(Received 1) December, 1, 1986;
2) December, 29, 1986)

Authors' addresses:

Dr. J. W. TOMM, Dipl.-Phys. B. SUMPF, Prof. Dr. sc. Karin HERRMANN
Humboldt-Universität zu Berlin, Sektion Physik
Invalidenstr. 110
Berlin
DDR-1040

Dr. A. SZCZERBAKOW, Instytut Fizyki, PAN, Warszawa, Poland

Cryst. Res. Technol.	22	1987	7	K113–K115

Short Notes

H. Jaskólska, L. Waliś, Cz. Janusz

Institute of Nuclear Chemistry and Engineering, Warsaw, Poland, and
Institute of Electronic Materials Technology, Warsaw, Poland

Investigation of Calcium Distribution in GGG Single Crystals by Means of Radiotracer Methods

Calcium presence in single crystals of $Gd_3Ga_5O_{12}$ (GGG) worsens their quality, so its concentration should be as low as possible. Therefore, gaining information about calcium behaviour during the single crystal growing process and calcium distribution in the crystal is of both theoretical and practical importance.

To examine this problem we used radiotracers and autoradiography methods. The tracer, ^{45}Ca, was created by irradiation 27 mg of calcium carbonate in the nuclear reactor at a thermal neutron flux of $5 \times 10^{13}\ n \cdot cm^{-2} \cdot s^{-1}$ during 4 weeks.

The nuclear data for this isotope are given in Table 1. The obtained tracer of the activity 11.5 MBq (310 μCi) was cooled for 8 days and then added to a charge stoichiometric oxide mixture. The single crystal was grown by the Czochralski method.

Table 1
Nuclear data of ^{45}Ca

Abundance of matrix isotope	cross section for activ.	half-life	E_β
2.06%	1.1 barn	165 d	245 keV

Process parameters were:
growth rate 5/mm/h
rotation rate 40 r/min

The obtained single crystal was cut according to the following scheme:

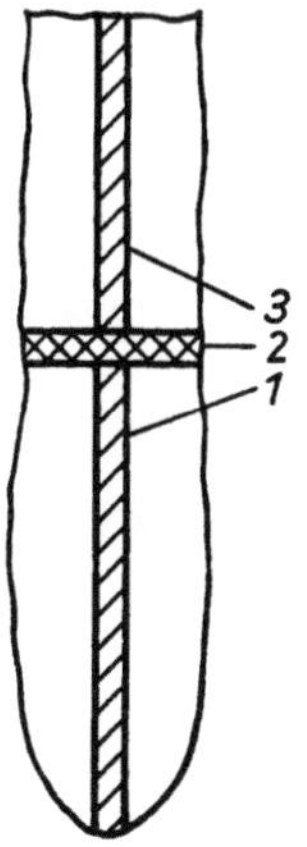

Fig. 1. Scheme of cutting the monocrystal

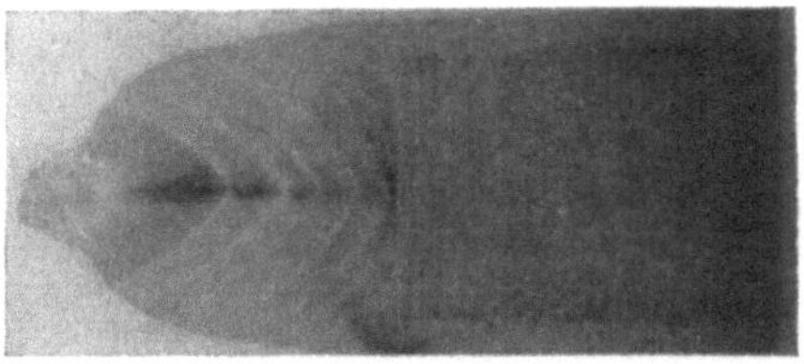

Fig. 2. Autoradiogram of the plate, cut out from the monocrystal along its vertical axis (a top part)

At the beginning of the process, significant periodic changes of temperature occur at the crystallization front. They are well visible in the first part of the crystal (Figs. 2, 3). Convex crystallization front is formed under the influence of the radial temperature gradient around the rotation axis of the crystal. The lowest temperature is at the crystal axis; so, the most intense deposition of calcium occurs there (local supercooling).

When the formation of the cone is finished and the crystal achieves its nominal diameter, the temperature changes become minimum. In consequence, the convex crystallization front changes its shape; it becomes flat (Fig. 2).

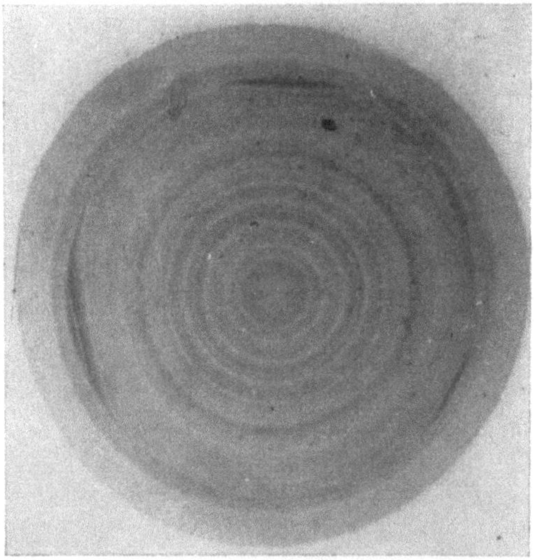

Fig. 3. Autoradiogram of the plate, cut out from the monocrystal perpendicularly to its vertical axis

Autoradiography revealed also the facet character of the crystal growth (Fig. 4). Alignment of facets in the outer layer differs from that in the remaining part of the crystal (Fig. 4). The proper selection of the crystal rotation rate, according to the given parameters of the single crystal growing process gives the correct flat shape of the crystallization front. Radiometric measurements indicate an increase in the calcium concentration along the axis of the single crystal (Tab. 2). This effect can easily

Fig. 4. Autoradiogram of the plate, cut out from the monocrystal along its vertical axis (an upper part)

Table 2
Relative specific activities of samples

Sample number	counting rate, $N/\frac{\text{counts}}{\text{min}}$	mass m/mg	N/m in a sample	N/m average
1	423	50.4	8.4	9.44
	555	52.9	10.5	
2	759	51.1	14.8	
	753	50.2	14.6	14.7
3	2302	49.9	46.1	45.8
	2285	50.3	45.4	
residue in a crucible	4361	51.2	85.2	82.0
	3952	50.2	78.7	

be explained, since the growth of the crystal is accompanied by an increase in calcium concentration in the material, remaining in the crucible.

The presented work shows the possibilities of using radioisotopic methods for studying the single crystal growing process.

(Received February 10, 1987)

Authors' addresses:

Dr. H. Jaskólska, Dr. L. Waliś, Asst. prof.
Institute of Nuclear Chemistry and Engineering
ul. Dorodna 16
03-195 Warsaw, Poland

Dr. Cz. Janusz
Institute of Electronic Materials Technology
ul. Wólczyńska 133
Warsaw, Poland

Cryst. Res. Technol.	22	1987	7	K116–K118

S. BHUSHAN, K. E. ABRAHAM

Department of Physics, Ravishankar University, Raipur, India

A Portable Electroluminescent Voltmeter

1. Introduction

Electroluminescence (EL) is the phenomenon of emission of light due to the application of electric field (a.c. or d.c.). The light emission strongly depends on the strength of the applied field. Thus, this property can be used to measure the voltage accurately. The first use of EL in the measurement of high voltage was made by SHAPOSHNIKOV and RAMAZANOV. The voltage measurement was based on a comparison of voltages by means of a capacitor with moving electrodes containing dielectric EL layer. The measurement was carried out up to 30 kV with high accuracy. BHUSHAN et al. used a simplified arrangement using a series of EL panels with different gaps. In both these voltmeters a suitable photomultiplier tube was used to detect the light intensity of EL emission. The present paper reports about a portable EL voltmeter which uses an EL panel and a photoconductive (PC) cell as a detecting unit.

2. Experimental

For the preparation of EL panel (ZnS, ZnO): Cu, Cl electroluminor with details of composition: ZnS 10 g, ZnO 2 g, (both luminescent grade), $CuSO_4$ 0.093 g, NH_4Cl 0.058 g and sulfur 0.2 g (all AR grade) fired at 1000 °C for 2.5 hrs was used. Araldite was used as a binder and the EL material was pressed between a conducting glass plate of resistance 150 Ω/cm^2 and an aluminium plate. A. C. voltages used for excitation were obtained from a variac coupled with a transformer. For detecting the light intensity of the El cell a PC cell using CdS was used. This PC cell was operated by a dry cell (9 V) and the photocurrent was recorded by a nanometer. As reported in this paper, the observed photocurrents are of the order of microampere and thus any micrometer could also be used. In this voltmeter the PC cell, the dry cell and the EL panel were enclosed in a black chamber and hence can easily be moved from one place to another.

3. Results and discussions

Figure 1 shows the schematic diagram of the experimental set up describing all the parts. Figure 2 shows the spectral curve of the EL material used, and thus the material used has been essentially the green emitting one. The photoconductor used has been CdS which has two peaks, one in the green region (edge emission) and the other in red region (extrinsic emission). The photoconducting rise and decay curves due to the EL light at two voltages of excitations are shown in Figure 3. Thus at lower voltages the rise and decay are very slow which become fast at higher voltages. The saturated photocurrent corresponding to different voltages of excitations are shown in Figure 4. The nature of photocurrent at different light intensities is responsible for this kind of

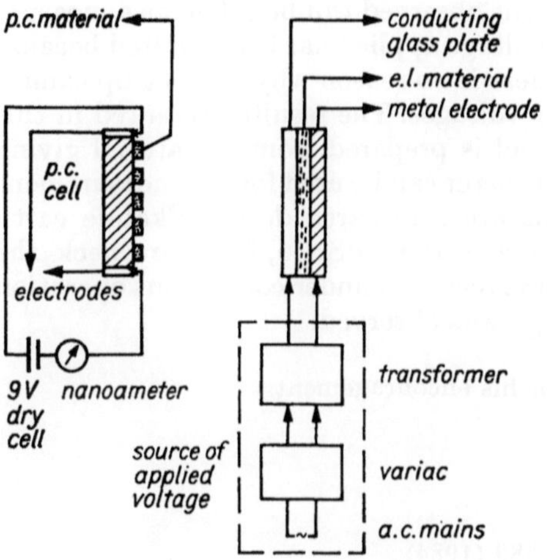

Fig. 1. Schematic diagram of the EL voltmeter

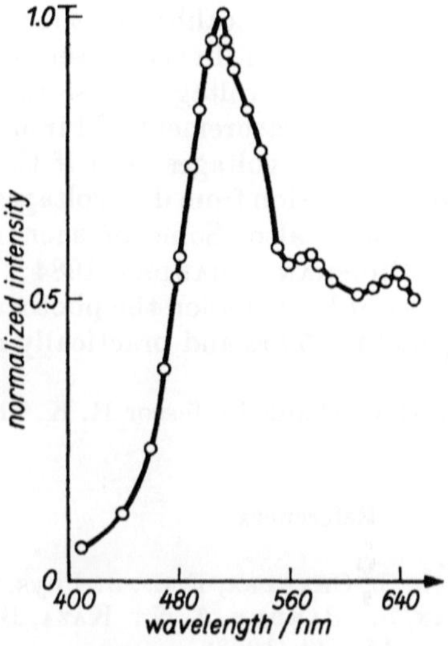

Fig. 2. Spectral curve of EL material under a.c. field (50 Hz) excitation

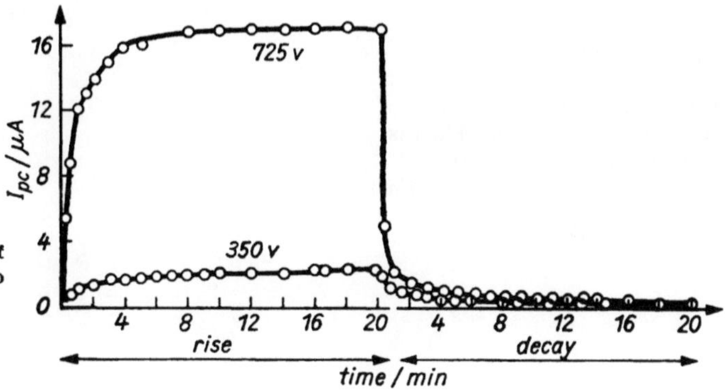

Fig. 3. Rise and decay curves of the photoconductor used at two excitation fields of EL panel

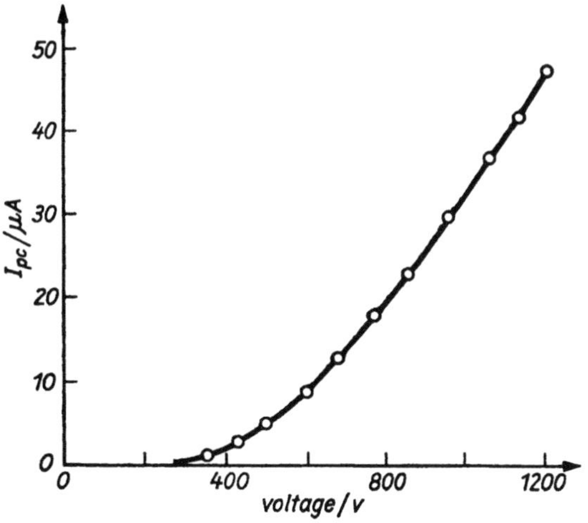

Fig. 4. Dependence of observed photocurrent on the applied a.c. voltages to the EL panel

behaviour. Thus once calibrated the photocurrent observed can be taken as a measure of the voltage applied. In the present case the voltage applied has been limited because of lack of very high voltage transformer but there is no reason why this set-up cannot be used for the measurement of further higher voltages. The results presented in this paper are for a.c. voltages, and if the EL panel is prepared from a material giving good steady emission from d.c. voltages the voltmeter can be used for the measurement of d.c. voltages also. Some of such materials are rare earth doped alkaline earth sulfides (BHUSHAN, CHANDRA 1984; VECHT et al.) and ZnS:Cu, Mn. To check the aging effect of EL emission the photocurrent was recorded under continuous operation of EL panel for 5 hrs and practically no change was observed.

The authors thank Professor R. K. Thakur for his encouragement.

References

BHUSHAN, S., CHANDRA, F. S.: J. Phys. (D) **17**, 589 (1984)
BHUSHAN, S., PANDEY, A. N., KAZA, B. R., SRIVASTAVA, L. K.: Ind. J. Pure & Appl. Phys. **15**, 744 (1977)
SHAPOSHNIKOV, A. N., RAMAZANOV, P. E.: Izv. Vuz. Fiz. **2**, 152 (1970)
VECHT, A., WAITE, M., HIGTON, M. H., ELLIS, R.: J. Lum. **24/25**, 917 (1981)

(Received February 11, 1987)

Authors' address:

c/o Dr. Shashi BHUSHAN
Suite No. 8, Teachers Hostes
Ravishankar University
Raipur, M. P., 492010, India

Cryst. Res. Technol.	22	1987	7	K119–K123

J. Rennie, M. A. S. Sweet

Robert Gordon's Institute of Technology, School of Physics, Aberdeen, England

Electroluminescence of Doped Calcium Sulphide Phosphors

A renewed interest in calcium sulphide is mainly due to the demonstration by Lehmann (1972) that the material can be doped by a wide range of dopants. This, combined with the large band gap (Lehmann 1970), means that a wide range of emitted colours can be produced (Lehmann 1972). Interest in electroluminescence centres on this possibility of a wider range of colours for display purposes then is available with existing commercial EL phosphors such as zinc sulphide.

To obtain EL from a powder sample a high electric field has to be applied. Since AC fields normally produce greater emission than DC fields alternating voltages were used in this preliminary work. The electrodes are placed close together with the phosphor between them, a given voltage producing a larger field for a thinner sample. The standard cell construction used in the work was a spring loaded device illustrated in Figure 1. A typical phosphor thickness of 100 μm was used. To allow detection of the luminescence one of the electrodes is made of conducting glass, the conducting layer on the glass usually being tin oxide. To prevent sparking it is necessary to exclude air from the gap between the electrodes. A binder is used for this purpose and this also serves to increase the light output from the CaS as it has an intermediate refractive index between that of CaS and air. This phosphor-binder layer can be produced by many methods and three of these will now be discussed.

The first method used was a spreading technique whereby the phosphor-binder mixture was spread to the required thickness using a chromatographic spreader. The thickness of the layer thus prepared depends on the height of the adjustable gate. The layer thickness was variable to a minimum of 200 μm. The binder material required for such a process is limited to those which will spread without forming agglomerations. Two binders which fulfilled this quality were liquid poly (Humbrol) and silicon grease. The liquid poly had the advantage that it was non-viscous. This property allowed homogeneous layers to be produced, however, these layers were

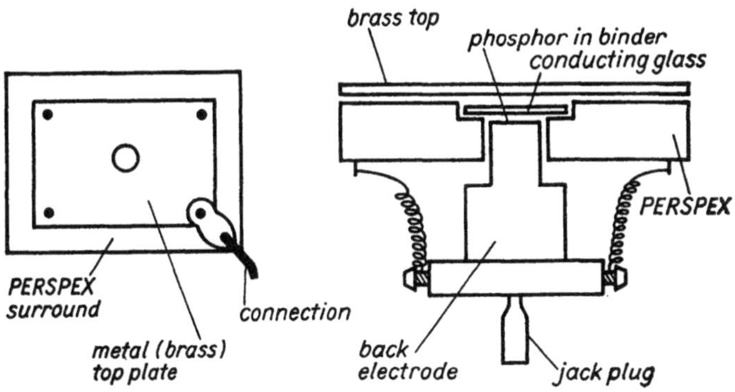

Fig. 1. Standard cell construction

fragile. Silicon grease gave well spread layers but remained in a colloidal solution with the result that the thickness tended to be altered when the layer was placed in the cell. However, the main difficulty with the solid layers was that they did not completely eliminate sparking in remaining air gaps between the boundaries of the electrodes and the layer.

The following settling technique was investigated. The conducting glass with an insulator attached (mylar) was placed in a solution of the powder and anyhydrous methanol. When the powder had settled out the glass was removed and the methanol allowed to evaporate off leaving the phosphor layer. Such layers were very fragile and sparking was still a problem. The major difficulty was that specified thicknesses could not be obtained and the method of varying the thickness was crude, i.e. varying the powder-methanol ratio.

The method which was the most effective employed a 'washer' type cell construction, see Figure 2. The conducting glass had a specially prepared mica washer attached

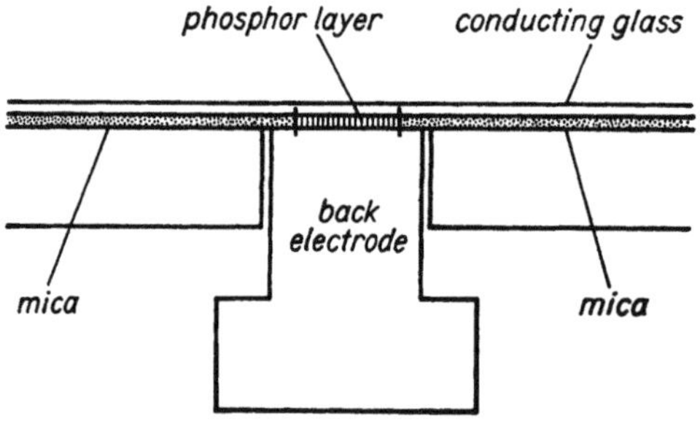

Fig. 2. Construction of the 'washer' type cell

to it. The diameter of this 'washer' was such that the back electrode of the cell mentioned above rested on the mica, the phosphor binder mixture being placed in the center of the 'washer'. This method enabled samples with thicknesses in the range 25 μm to 300 μm to be prepared. The thinner mica 'washers' being used the most due to the greater EL intensity obtained. The main binders used here were silicon grease and olive oil. It was found easier to produce mixtures using the olive oil but the layers produced formed air bubbles after the period of a day. This led to the results being unreproducible, furthermore such a cell is clearly not of permanent use.

The CaS samples which showed EL were prepared by a reduction reaction in which calcium sulphate is converted to calcium sulphide. To obtain the doping of the samples the relevant dopant concentrations were added to the sulphate which was then fired to give the doped calcium sulphide sample. All dopant levels quoted in this paper

Table 1

Comparison of the electroluminescence output of differently doped calcium sulphide phosphors. Excitation of 350 Vrms at a frequency of 10 kHz

Dopant	concentration of dopant (ppm)	binder	method of producing layer	relative brightness
Ce, Cu	400	—	evaporation	3.16
Er, Cu	400, 11000	silicon grease	spreader technique	4.17
Pr, Cu	400, 4000	olive oil	'washer' cell	2.10
Cu, Cl	300	olive oil	'washer' cell	14.38
Al	100	olive oil	'washer' cell	1.75
Cu	100	olive oil	'washer' cell	2.40

refer to the levels added to the unfired samples. The firing of the samples took place in a silicon tube furance at 1000 °C with hydrogen gas flowing, the time of firing being three to four hours. Subsequent to firing the samples were slowly (over 8 hours) cooled to room temperature in a hydrogen atmosphere.

A number of dopant materials was investigated and in Table 1 the relative intensities of emission of some of those which produced EL are listed along with the phosphor layer preparation technique used in each case. The measurement of intensity was carried out by a Thorn-EMI photomultiplier model 9804B. Typical brightness voltage curves are shown in Figures 3 and 4. These curves, from a CaS:Cu, Cl sample,

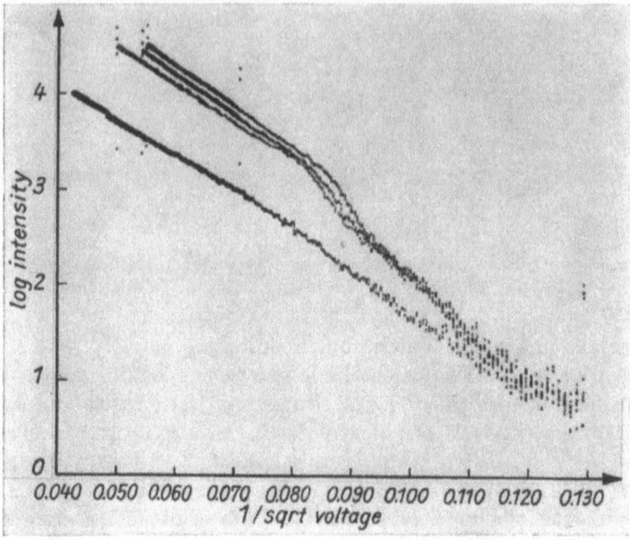

Fig. 3. Plots of log (intensity) against log (voltage) at various excitation frequencies (from top to bottom 10, 6, 2, and 0.8 kHz) for the sample CaS/Cu, Cl (300 ppm)

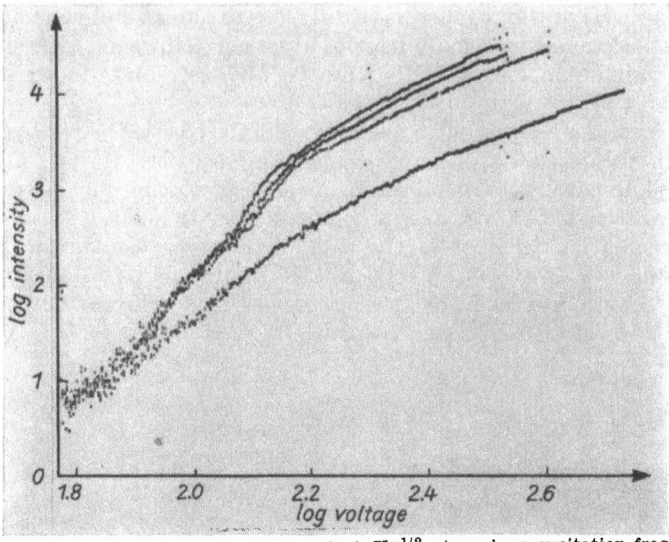

Fig. 4. Plots of log (intensity) against $V^{-1/2}$ at various excitation frequencies (from top to bottom 10, 6, 2, and 0.8 kHz) for the sample CaS/Cu, Cl (300 ppm)

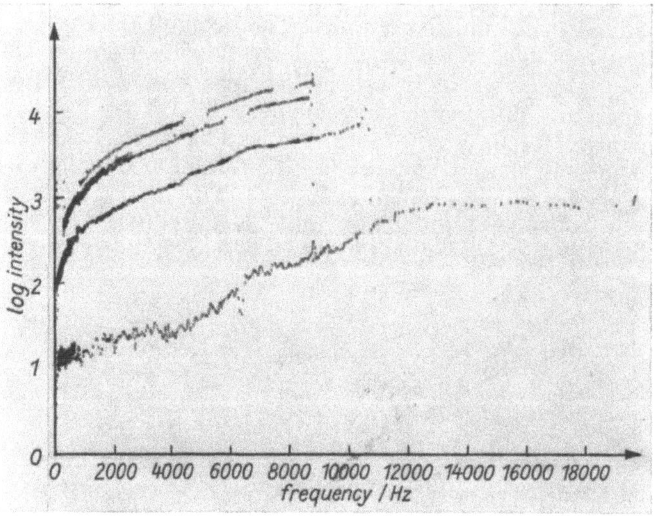

Fig. 5. Plots of log (intensity) against frequency at various excitation voltages (for top to bottom 400, 350, 300 and 100 V_{RMS}) for the sample CaS/Cu, Cl (300 ppm)

show the two main relationships to which such results are usually fitted. Brightness frequency curves are plotted in Figure 5 which shows the basic trend obtained for all the samples examined. These graphs are consistent with those published in the literature on CaS phosphors (Lawangar et al.; Patil, Lawangar; Wachtel; Rastogi, Mor) and are also very similar to those published on ZnS (Ivey; Gumlich). The voltage results Figures 3 and 4 perhaps favour the inverse square root relationship since this is the graph with the greater linearity. This is often interpreted as proof of the operation of an acceleration-collision excitation mechanism (Ivey; Alfrey, Taylor; Zalm). However, it is unwise to draw such a conclusion from such a complex emission. The emission is complex due to a multitude of independently emitting regions in the sample all of which must have differing variables and constants. A study of one emitting 'spot' would be a more fruitful approach. However, there are difficulties in isolating a single spot exacerbated by the high refractive index of CaS, and also of working with very low intensity signals.

Although most samples examined to date have exhibited EL the emission from the CaS phosphors investigated has been disappointing because of low intensity and patchy emission. The patchy emission indicates that the techniques used to prepare the powders are far from optimised. From the results to date it can be seen that copper is an important impurity as most of the samples showing significant EL contain copper. Thus future work will concentrate on the sample preparation techniques to optimise EL emission and extend the spectral range. This will include examinations of single crystals and thin films as well as powders.

References

Alfrey, G. F., Taylor, J. B.: Proc. Phys. Soc. (London) **B68**, 775 (1955)
Gumlich, H. E.: J. Lumin. **23**, 73 (1981)
Ivey, H. F.: Electroluminescence and Related Effects, London 1963
Lawangar, R. D., Pawar, S. H., Narlikar, A. V.: Mat. Res. Bull. **12**, 341 (1977)

LEHMANN, W.: J. Electrochem. Soc. **117**, 11, 1389 (1970)
LEHMANN, W.: L. Lumin. **5**, 87 (1972)
PATIL, M. G., LAWANGER, R. D.: Indian J. Pure Appl. Phys. **20**, 139 (1982)/
RASTOGI, A. M., MOR, S. L.: Indian J. Pure Appl. Phys. **14**, 187 (1976)
WACHTEL, A.: J. Electrochem. Soc. **107**, 3, 119 (1960)
ZALM, P.: Philips Res. Rep. **11**, 353 (1956)

(Received February 26, 1987)

Authors' address:

c/o Dr. M. A. S. SWEET
Robert Gordon's Institute of Technology, School of Physics
St. Andrew Street
Aberdeen AB1 1HG, England

Crystal Research and Technology

Zeitschrift für experimentelle und technische Kristallographie

Edited by H. NEELS, Leipzig

Papers in Number 8, 1987

Original Papers

AKADEMIE-VERLAG · BERLIN

Continued from cover page 4

Short Notes

Book Reviews

Papers im Number 8, 1987

Short Notes

Crystal Research and Technology
Volume 22 1987 Number 7

Contents

Original Papers

Continued on cover page 3

Crystal Research and Technology is indexed in Current Contents/PC & ES.

Bei Fragen zur Produktsicherheit wenden Sie sich bitte an:
If you have any questions regarding product safety,
please contact:

Walter de Gruyter GmbH
Genthiner Straße 13
10785 Berlin
productsafety@degruyterbrill.com